Multi-ring basins are large impact craters formed in the early history of planets. They critically affect the evolution of the planets and satellites. The Moon offers an exceptional chance to study these phenomena and this book provides a comprehensive geological study using data from lunar landings and remote sensing of the Moon.

The author covers the formation and development of basins and considers their chemistry and mineralogy. He studies their effect on the volcanic, tectonic and geological evolution of the planet, including the catastrophic consequence on the planetary climate and evolution of life. The terrestrial planets are also examined. This study is lavishly illustrated with many spectacular, highly detailed photographs and diagrams.

Cambridge Planetary Science Series

The
Geology
of
Multi-Ring
Impact Basins

Cambridge Planetary Science Series

EDITORS: W. I. Axford, G. E. Hunt, and R. Greeley

The
Geology
of
Multi-Ring
Impact Basins

The Moon
and
Other Planets

The
Geology
of
Multi-Ring
Impact Basins

The Moon
and
Other Planets

Lunar and Planetary Institute
Houston, Texas

PUBLISHED BY THE PRESS SYNDICATE OF THE UNIVERSITY OF CAMBRIDGE
The Pitt Building, Trumpington Street, Cambridge, United Kingdom

CAMBRIDGE UNIVERSITY PRESS
The Edinburgh Building, Cambridge CB2 2RU, UK
40 West 20th Street, New York NY 10011–4211, USA
477 Williamstown Road, Port Melbourne, VIC 3207, Australia
Ruiz de Alarcón 13, 28014 Madrid, Spain
Dock House, The Waterfront, Cape Town 8001, South Africa

http://www.cambridge.org

First published 1993
First paperback edition 2005

A catalogue record for this book is available from the British Library

Library of Congress cataloguing in publication data

Spudis, Paul D.
The geology of multi-ring impact basins: the Moon and other
planets / Paul D. Spudis.
p.   cm. – (Cambridge planetary science series; 8)
Includes bibliographical references.
ISBN 0 521 26103 1 hardback
1. Multiring basins (Astrogeology) 2. Lunar basins. I. Title.
II. Series.
QB456.S64   1993
551.3′97′09991–dc20   93-3236 CIP

ISBN 0 521 26103 1 hardback
ISBN 0 521 61923 8 paperback

Contents

Preface

Multi-ring basins are the largest impact craters on Solar System bodies. They form in the earliest stages of planetary history by the collision of asteroid-sized bodies with planets and affect the subsequent evolution of these latter objects in many profound ways. Many scientists have expended great effort in attempting to understand these features; a casual glance at the literature of planetary science over the last 30 years reveals no less than several hundred entries dealing with some facet of multi-ring basins.

Planetary scientists studying the problem of multi-ring basins approach it from many different directions. Some are physicists, describing the mechanics of basin formation on the basis of known theory. Other workers make geological maps from photographs, searching for clues to the processes that have shaped the surface of the planet. Still others study the chemistry and mineralogy of terrestrial and lunar samples, using the rock record to reconstruct the physical extremes of heat and pressure produced during large impacts. The basin problem is multi-disciplinary; answers to the many questions raised by these features require knowledge from geology, chemistry, physics, and other fields of study. No one person has the expertise to understand all aspects of the basin problem: So why this book?

The only other book available on the problems posed by basins is the proceedings of a topical conference held at the Lunar and Planetary Institute, Houston, in November, 1980 (*Multi-ring Basins: Formation and Evolution*, P.H. Schultz and R.B. Merrill, editors, Supplement 15 of *Geochimica et Cosmochimica Acta*, Pergamon Press, New York, 1981). This volume contains 20 papers by separate authors dealing with a variety of sub-topics of the multi-ring basin problem. Since that time, although several books dealing with lunar science and impact cratering have appeared, no single work dealing with the geology of multi-ring basins on the planets has been published.

As a geologist, I am interested in the processes that have shaped the planets and their histories. Basins are the fundamental causes of the current configuration of some planetary surfaces; I believe that the geological approach to the basin problem can make a significant contribution towards its solution. Geological studies cannot solve the basin problem in its entirety, but it was a geologist who first recognized the impact origin of the Imbrium basin on the Moon (G.K. Gilbert, 1893) and uncounted geologists since Gilbert's pioneering study have mapped, classified, and contemplated these features on all the planets.

The central thesis of this book is that of all the terrestrial planets, Earth's Moon is the best place in the Solar System to study multi-ring basins. The Moon is a relatively primitive object, largely unmodified for the last three billion years. Moreover, the Moon has had a complex igneous history; it was recognized very early that although basins are impact-generated landforms, they are intimately related to planetary igneous processes (e.g., most volcanic plains on the planets tend to fill basins). The Moon, being relatively close, was the first planetary body to be explored since the Space Age began. It has been the target of flybys, orbital missions, hard and soft landers, sample returns, and human exploration. Because of this large and excellent data base, we can address basin problems on the Moon to a degree unsurpassed on any other planet.

I have chosen to focus on five basins (discussed in Chapters 3–7 of this book) that typify the range of features associated with multi-ring basins and that have the best coverage of all data types. No basin is covered by all sources of data as well as we would like, but thanks to the legacy of the Apollo program, a wealth of information is available and can be used to understand such problems as the mechanics of basin formation, impact melting and emplacement of ejecta, modification of basin topography and rings, and the long-term effects of basins on the geological evolution of the planets. I do not pretend to be an expert on all these aspects of basin studies, but rather, offer this book in the belief that a synthesis of a variety of information using the geological approach provides some important constraints that must be satisfied to solve the multi-ring basin problem. The reader will notice that many questions raised in this book have yet to receive satisfactory answers; there is still scientific gold to be mined from the study of planetary basins.

My work on impact basins has been and is supported by the Planetary Geology and Geophysics Program of the National Aeronautics and Space Administration and I gratefully acknowledge this continuing support. I thank the Branch of Astrogeology of the U.S. Geological Survey in Flagstaff, Arizona, for providing me with both intellectual stimulation and a base for my research for over ten years. Portions of this work were performed during visits to the facilities of the Planetary Geosciences Group, University of Hawaii, Honolulu, the University of London Observatory, London, and the Center for Earth and Planetary Studies at the National Air and Space Museum, Washington, and this support is acknowledged with thanks.

One of the pleasures of scientific research is working with dedicated colleagues; I have been particularly fortunate to have collaborated with some scientists that have been both intellectually stimulating taskmasters and also, good friends. Ron Greeley, more than any other person, is responsible for my career, first as my employer and then as my dissertation advisor at Arizona State University. As the Science Editor for the Cambridge Planetary Science Series, Ron has also consistently encouraged me to finish this book and I hope that the final product justifies his efforts. Don Wilhelms, late of the U.S. Geological Survey, and I have spent many hours debating the issues raised by multi-ring basins; while he and I have agreed to disagree on several points, I want to express my admiration and respect for Don as one of the premier lunar scientists in this business. His contributions to unraveling the surprisingly complex geological history of the Moon are unsurpassed by the many who have worked on this problem.

Over the years, it has been my fortune and pleasure to collaborate on a variety of research projects with Pete Schultz, B. Ray Hawke, Graham Ryder, John Guest, Phil Davis, Dick Pike, Mark Cintala, Richard Grieve, and Jeff Taylor; these scientists have always been a pleasure to work with and constantly provide me with food for thought. I thank Odette James, Dick Pike, and Buck Sharpton for providing illustrations. Dick Pike, Jeff Taylor, and Graham Ryder undertook and completed a truly herculean task: that of reviewing all or parts of this book. I thank them for their considerable efforts and also absolve them of responsibility for errors of fact or interpretation presented in these pages.

1

The multi-ring basin problem

Ever since their recognition, multi-ring basins have fascinated and vexed scientists attempting to reconstruct the geological history of the Moon. As the other terrestrial planets were photographed at high resolution, it became apparent that basins are an important element in the early development of all planetary crusts. This importance spurred research into the basin-forming process and yielded a plethora of models and concepts regarding basin origin and evolution. In this chapter, I outline the general problem areas of basin formation and describe the approach taken by my own work on lunar basins.

1.1 Multi-ring basins and their significance

Multi-ring basins are large impact craters. The exact size at which impact features cease to be "craters" and become "basins" is not clear; traditionally, craters on the Moon larger than about 300 km have been called basins (Hartmann and Wood, 1971; Wilhelms, 1987). Basins are defined here as naturally occurring, large, complex impact craters that initially possessed multiple-ring morphology. This definition purposely excludes simulated, multi-ring structures that result from explosion-crater experiments on the Earth and whose mechanics of formation differ from impact events (e.g., Roddy, 1977), although important insights into the mechanics of ring formation may be gained from these studies. The qualification that basins *initially* possessed multiple rings is in recognition of the fact that many older, degraded basins display only one or two rings, even though their diameters of hundreds of kilometers indicate that they had multiple rings when they originally formed.

The distinction should also be made in the case of lunar basins between the terms *basin* and *mare*. Basin refers to the circular structure produced by the collision of a large planetesimal (asteroid) with the Moon; such features may be subsequently flooded by volcanic materials, but many are only partly filled or are not filled at all. The term mare is used to refer to any dark, plains-forming material and may or may not be contained exclusively in a basin. Thus, Mare Imbrium refers to only the low-albedo plains that mostly fill the Imbrium basin.

Basins are ubiquitous on the terrestrial planets and typically occur within the most ancient, heavily cratered terrain on the Moon, Mars, Earth, Mercury, and some of the icy satellites of Jupiter and Saturn. Multi-ring basins formed in the earliest stages of terrestrial planet evolution and exerted important influence on the subsequent geological history of these bodies. Moreover, the extensive structural effects on planetary lithospheres produced by a basin impact may alter the spatial distribution of ongoing geological processes, in particular, the timing and distribution of volcanism. Basin sites frequently act as loci for the eruption of volcanic materials and this relation was responsible for the confusion between the terms basin and mare in some early studies. The crustal unloading resulting from a basin impact may have far-reaching consequences, as long-term rebound and crustal adjustment can alter the location of subsequent planetary volcanism and tectonism.

A basin-forming impact extensively redistributes vast amounts of crustal materials, thereby changing the regional chemistry and petrology of planetary crusts. Samples returned by spacecraft missions are typically collected from surficial deposits. On the Moon, and possibly samples returned from Mars in the future, some of these samples may be related to basin deposits. Thus, it behooves us to understand better the many effects of basin formation on the planetary surface compositions and the emplacement of ejecta.

Among terrestrial planets, the Moon is a particularly suitable object for the study of multi-ring basin problems. As a relatively primitive body in the geological sense, many of the Moon's earliest features are well preserved. In addition, the Moon has been extensively photographed, both from the Earth and from spacecraft, at a variety of viewing geometries and lighting conditions. These photographs permit clear

Figure 1.1 Index map of the Moon (Lambert equal-area projection), showing the location of five multi-ring basins whose geology is analyzed in detail in Chapters 3–7 of this book.

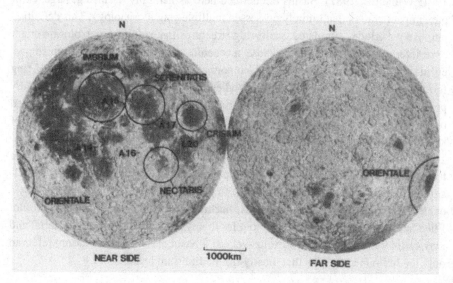

delineation of morphologic features associated with basin deposits and structures. Remote-sensing techniques furnish geochemical, mineralogical, and geophysical data for lunar basins which allow us to infer the regional chemistry and petrology of basin ejecta deposits, local variations in crustal thickness, and basin topographic and gravimetric properties. Finally, the American Apollo and Soviet Luna missions obtained samples of basin deposits and much detailed work on these samples has revealed the chemical and petrologic properties of the lunar crust and has provided tantalizing clues to the impact process.

A detailed study of multi-ring basins synthesizing all of this information has never been attempted. I have studied five basins on the Moon (Figure 1.1), collating multiple and diverse sets of data. My objective is to integrate these data into a coherent geological model for basin formation and evolution that not only sheds light on processes involved in lunar basin evolution, but also produces constraints on the general problem of multi-ring basin formation on all of the terrestrial planets.

1.2 Overview of the lunar multi-ring basin controversy

Although intensively studied, the problems surrounding the formation and evolution of multi-ring basins have been responsible for much debate, but little consensus. The controversy over basins largely revolves around questions of the size and shape of the crater of excavation and the origin of multiple rings. By way of acquainting the reader with the debate over these issues, I here review both the history of basin studies and some of the key questions associated with their genesis.

1.2.1 Recognition of multi-ring basins

When Galileo first studied the Moon through the telescope in 1609, he saw that large, overlapping craters make up the lunar highlands; he also sketched the arcuate mountain ranges on the Moon. Galileo and subsequent workers recognized that the mountains bordering the circular maria are similar to crater rims, a key link towards understanding that basins are merely large craters and have the same origin. Later workers mapped the Moon's surface in increasing detail, usually with the aim of advocating some mechanism of crater formation, typically, as a result of volcanism. Few scientists proposed that impact was an important process in crater genesis, but Gruithuisen (1829) and especially Proctor (1873) championed the impact origin of craters on the Moon.

The first study to advocate an impact origin for lunar basins was the classic paper of Grove Karl Gilbert (1893), who was at the time Chief Geologist of the U.S. Geological Survey. Although Gilbert was primarily concerned with the origin of lunar craters, he was the first to recognize the extensive pattern of "sculpture" surrounding the Imbrium region of the Moon and hypothesized that this was produced by a flow of "pasty" debris that proceeded outward in radial directions from the Imbrium "collision" area, a surprisingly modern view (see review by Oberbeck,

1975). To buttress his conclusions that lunar craters formed by impact, Gilbert studied Meteor crater, Arizona (Gilbert, 1896), ironically concluding that this feature resulted from a volcanic steam explosion! (For a detailed and fascinating history of the study of Meteor crater and the development of impact theory, see Hoyt, 1987).

Historiographical myth has it that Gilbert's ideas on the Moon were ignored because he chose to publish his work in the *Bulletin* of the Philosophical Society of Washington, a journal rarely consulted by today's lunar scientists. In fact, Gilbert was one of the most eminent men of science in the nineteenth century and his opinions regarding the origin of lunar craters were bound to attract notice (see Hoyt, 1987, pp. 65–67). The reluctance of Gilbert's scientific peers to embrace impact as an important process largely resulted from the lack of a clear example of an impact crater on the Earth; surely a process of such importance should be detectable somewhere! Thus, the few that did think about the Moon preferred to interpret its surface largely in terms of processes (such as volcanism) that made holes and that were well displayed and (allegedly) well understood on the Earth. In this intellectual milieu, Gilbert's work on the Moon was forgotten. A contributory factor to this relative obscurity was his own reluctant pronouncement on the volcanic origin of Meteor crater. Gilbert's vision of the geology of the Moon had to await the space age to be read and appreciated.

Astronomers of the early twentieth century tended to favor volcanic origins for the surface features of the Moon (see Hoyt, 1987). Few geologists thought much about either the Moon or the process of impact, but two landmark papers are noteworthy. Boon and Albritton (1936) suggested that meteorite impact was responsible for the origin of a variety of terrestrial features known as "cryptovolcanic structures". The recognition of a class of impact craters with diameters of tens of kilometers on the Earth permitted detailed investigation of the chemical and physical processes at work in very energetic impacts. Also, an important contribution by Dietz (1946) outlined a new geological sketch of the Moon similar to that of Gilbert (1893), involving impact origins for craters as well as the "maria" (i.e., basins); like Gilbert, Dietz believed that the dark mare plains filling the Imbrium depression were generated during the basin-forming impact and he considered volcanism on the Moon to be possible, but of minor significance.

Two important twentieth century works by Baldwin (1949, 1963) extended the discussion of impact basins to include all of the circular maria on the near side of the Moon. Baldwin (1949) presented a model for the geology of the Moon that was nearly completely accurate, concluding that most craters formed very early in the lunar history by impact of asteroidal objects and that the dark maria are flows of basaltic lava, unrelated genetically to the basins that contain them. In this and his subsequent work, Baldwin (1949, 1963) used the term "circular maria" to refer to basins because the basins on the near side of the Moon are nearly always completely filled by mare basalt. This terminology produced confusion in the minds of many workers (e.g., Urey, 1952), who conflated the topographic features produced by impact (basin) and the unrelated, subsequent fill by volcanic lava flows (maria).

In the course of mapping the geology of the Moon, Shoemaker and Hackman (1962) clearly marshaled the evidence that the basins formed substantially before the surface flows of basaltic maria; they also developed a simple and practical stratigraphic system for the Moon based on the use of ejecta deposits of basins as marker beds. This system permitted the relative ages for basins and other surface features to be delineated. As systematic mapping of the geology of the Moon continued, many new basins were discovered and new appreciation emerged for the control by basins of the basic structural framework of the lunar surface (e.g., Stuart-Alexander and Howard, 1970).

In the same years that Shoemaker and Hackman were developing the rationale for the geological mapping of the Moon, a project at the Lunar and Planetary Laboratory at the University of Arizona was to lead to a new dimension in basin studies. By projecting transparencies of Earth-based telescopic photographs of the Moon onto a blank sphere, William K. Hartmann produced a series of geometrically rectified views of the lunar basins. This led to the recognition that the basins were surrounded by a series of multiple, concentric *rings*. During systematic study of the near side of the Moon, Hartmann identified multiple rings associated with all of the mare basins (Hartmann and Kuiper, 1962; for a personal and excellent reminiscence on these discoveries, see Hartmann, 1981).

The systematic pre-Apollo geological mapping program of the U.S. Geological Survey, summarized in Wilhelms (1970) and in Stuart-Alexander and Howard (1970), surveyed the entire Moon, demonstrated the prevalence of basins in its early evolution, and documented the profound influence of basins on regional geological patterns. Continued study of basins led to debate concerning the particulars of their formation. The debate focused mainly on two different, but closely related questions: (1) What were the dimensions of the original crater of excavation for these large impact features? and (2) How do multiple rings form? The next sections summarize these controversies.

1.2.2 The problem of the original crater of excavation

It was recognized early that although basins are very large impact craters, the profound differences in morphology between basins and smaller craters must be a reflection of fundamentally different processes at work during and after the impact event. A major problem of basin geology is the identification of the original crater of the basin or *transient cavity* (Table 1.1). This identification is required for understanding the mechanics of basin formation, crustal volume displacement, lunar sample provenance, and the origin of *mascons* (anomalous concentrations of mass in the lunar surface).

One of the earliest hypotheses for the size of the original craters of basins resulted from studies of the mechanics of formation of smaller impact craters, both natural and experimental. These studies suggested that the initial crater for lunar basins was a feature much smaller than the currently expressed *topographic rim*

Table 1.1 *Concepts and models for the excavation of multi-ring basins*

Models	
Group A	Group B
Basins are scaled-up versions of smaller impact craters (proportional growth)	Basins form by fundamentally different cratering mechanisms (non-proportional growth)

Proponents	
Gault, 1974	Baldwin 1974, 1981
Croft, 1980, 1981a, 1985	Hodges and Wilhelms, 1978
Grieve *et al.*, 1981	Murray, 1980
Schultz *et al.*, 1981	Wilhelms, 1984, 1987
Spudis, 1986	Schultz and Gault, 1986
Spudis *et al.*, 1984, 1988b, 1989	

Arguments	
For	
(1) Predicts excavated depths and volumes that agree with sample and geophysical data	(1) Predicts shallow excavation cavities, in agreement with sample data
(2) Does not invoke hypothetical and untestable mechanisms	(2) Explains basin inner rings by either strength differences or oscillatory uplift and collapse
Against	
(1) Experimental evidence indicates that flow fields might change with increasing crater size	(1) Some mechanism changes cratering process above a certain size limit (*deus ex machina*)
(2) Does not explain basin rings; these must form during modification stage	(2) Does not explain basin *outer* rings

Source: Adapted from Pike and Spudis (1987)

(Figure 1.2; Gault, 1974; Dence, 1976, 1977a; Schultz, 1976a). A tradition established early in basin studies was the association of one of the observed rings with the *excavation cavity*, the feature from which material is ejected during the formation of the basin. Most of the studies mentioned above equated one of the innermost rings with the boundary of excavation during basin formation. In contrast, Schultz (1979) suggests that for some lunar basins, particularly the older ones, post-impact structural modification has completely obliterated any morphologic evidence for the original crater rim, but the boundary of the transient cavity still lies well within the current basin rim.

The depth to which these small cavities excavated is controversial; the use of terrestrial analogs (Dence *et al.*, 1974; Dence, 1976) initially suggested deep excavation, with depth–diameter ratios approaching those of simple craters (approximately 1:3). Croft (1980, 1981a) suggested that the use of a simple cratering flow model

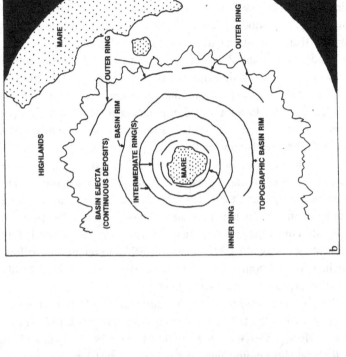

Figure 1.2 Photograph (a), sketch map (b), and hypothetical cross section (c) of a typical multi-ring basin on the Moon (Orientale), showing specific features discussed in this book. For more details on the Orientale basin, see Chapter 3.

may better approximate the effective excavation depth for small cavity basin formation. The "Z-model" of Maxwell (1977) attempts to explain cratering phenomena by analysis of hydrodynamic flow of particles during the crater excavation phase. In this model, particles follow flow streamlines radiating from a point source within the transient cavity. In Z-model results, effective excavation by the crater is from depths shallower than that of the transient cavity (Maxwell, 1977; Croft, 1981a). For basin-sized impacts, the Z-model suggests a depth–diameter ratio of the excavation cavity of about 1:10, a significant reduction of Dence's 1:3 estimate.

Many workers have relied on morphologic and stratigraphic relations seen in lunar photographs to estimate the location of the original crater. Most of these studies suggest that a ring within the main rim of the basin, usually at some intermediate position (Figure 1.2), best approximates the location of the transient cavity of the basin (Head, 1974a; Moore *et al.*, 1974; McCauley, 1977; Scott *et al.*, 1977). In this view, the innermost rings of basins represent central uplifts, analogous to central peaks seen in smaller craters, and the outer rings form in response to structural adjustment of the crust to an initially smaller crater. The estimates of the depths of these transient cavity structures vary widely, ranging from very deep, implied by the simple crater geometry assumed by Moore *et al.* (1974), to relatively shallow, resulting from the increasing dominance of substrate strength, which allegedly would reduce excavation depth by substantial amounts (Head *et al.*, 1975).

Other workers, studying the same photographs, interpret the main topographic rim of the basin (Figure 1.2) as the boundary of excavation (Baldwin, 1974, 1981; Wilhelms *et al.*, 1977; Hodges and Wilhelms, 1978; Murray, 1980). In this view, basins are simply the scaled-up equivalents of the rims of smaller lunar craters, which these authors believe represent the original cavity of excavation, enlarged little by slumping. Opinions differ, however, as to the nature of the cratering process. Baldwin and Murray believe that basins are true analogs to smaller craters, whereas Wilhelms and co-workers advocated a "nested crater" model which contends that basin interior rings are the result of strength-dependent layering in the lunar crust. In this view, basin rings are analogous to sub-kilometer-sized concentric craters formed in the maria, where unconsolidated debris overlies coherent basaltic bedrock (Quaide and Oberbeck, 1968). In both of these models, the depths to which basin impacts may excavate is an order of magnitude greater than estimates from models that identify the transient cavity rim with one of the inner basin rings. Finally and most recently, Pike and Spudis (1987) specifically repudiate the concept that *any* of the currently expressed rings *must* represent the original crater of the basin, noting that such an equivalence merely is assumed and is not required to model or understand the formation of basins.

1.2.3 *The origin of basin rings*

Understanding the origin of basin rings is dependent in part upon the identification of the original crater rim. In general, those workers favoring a small excavation crater favor structural origins for ring systems; those favoring large initial craters believe that physical properties of the lunar crust, either during or after the impact, are responsible for rings (Table 1.2). These two modes of origin are not entirely mutually exclusive and some models incorporate features of both.

The concept of a *megaterrace*, a large concentric slump feature around a small excavated crater, was one of the first ideas to be advanced for the origin of basin rings (Hartmann and Kuiper, 1962; Mackin, 1969; Gault, 1974; Dence, 1976, 1977a; Head, 1974a, 1977a; McCauley, 1977; Melosh and McKinnon, 1978). All of the above workers agree on this origin for the main topographic rim of basins (Figure 1.2); both Gault and Dence visualize all of the basin outer rings as being produced by this mechanism around a small, simple transient cavity. Those authors equating an intermediate ring (Figure 1.2) with the basin cavity suggest a dual mechanism whereby the main rim is a megaterrace, but the inner rings are rebound features, analogous to central peaks seen in smaller craters (Head, 1974a; McCauley, 1977; Scott *et al.*, 1977). The megaterrace hypothesis is based in part on analogous slumping evident in smaller lunar craters (see Chapter 2). This slumping in basin-forming events cannot be scaled up from complex craters directly; it requires that the lunar crust respond as if it consisted of coherent blocks during the modification stage of basin formation.

A completely different approach to ring genesis is taken by Baldwin (1974, 1981) and Murray (1980). They argue that an impact large enough to produce a basin would "fluidize" the lunar crust in the target region and that wave-like phenomena would produce rings like the ripples in a pond after a stone had been dropped into it. The use of the term "fluidization" is used here in the rheological sense that lunar crustal materials behave in a fluid manner (Melosh, 1979); this use does not imply any volatile content within crustal materials on the Moon. Van Dorn (1968) and Baldwin (1974) suggest that lunar "tsunamis" produced by the impact of a basin-forming projectile would freeze in place after impact excavation, producing basin rings. Murray (1980) envisions an oscillatory mechanism whereby "fluidized" lunar crust would produce ring systems by continuous overthrust of ripples until the crust had solidified into a multi-ring plan.

Yet another mechanism for ring production requires target strength to be the dominant factor in ring formation. An example of such a mechanism is the nested crater model advocated by Wilhelms *et al.* (1977) and Hodges and Wilhelms (1978). In this model, basin interior rings reflect the presence of global layering in the lunar crust, each layer having different strengths. Basin rings represent the boundary of excavation of a series of nested craters into deeper crustal levels; the basin topographic rim marks the boundary of excavation. Melosh and McKinnon (1978), McKinnon (1981), and Melosh (1989), while favoring a dominantly structural origin for basin rings, suggest that rings can form only under conditions in which the planetary lithosphere (rigid outer layer) is of such a thickness as to permit structural failure. Thus, this model also falls into the "target strength" category of ring-forming mechanisms (Table 1.2).

Table 1.2 *Concepts and models for basin ring origin*

Models

Group A	Group B	Group C
Impact-driven wave mechanism; dominantly surficial	Deep-seated structures, modification phase; "megaterraces"	Target strength or thickness differences; syn- or post-impact

Proponents

Baldwin, 1974, 1981	Hartmann and Wood, 1971	Melosh and McKinnon, 1978
Murray, 1980	Head, 1974a, 1977a	
Pike, 1983	Howard *et al.*, 1974	Hodges and Wilhelms, 1978
Van Dorn, 1968	Mackin, 1969	
Chadderton *et al.*, 1969	McCauley, 1968	McKinnon, 1981
		Melosh, 1989

Grieve *et al.*, 1981
Croft, 1981b
Pike and Spudis, 1987

Arguments

For

(1) Evidence for oscillatory uplift at terrestrial impact craters	(1) Geologic evidence for structure associated with rings	(1) Evidence for strength-related excavation at Ries crater
(2) Potentially explains √2 *D* ring spacing	(2) Gravity-driven collapse observed at complex craters	(2) Lithospheric thickness an overlooked variable

Against

(1) Physical plausibility uncertain	(1) Does not explain √2 *D* spacing or outer rings	(1) Does not explain √2 spacing or outer rings
(2) Evidence for deep structures associated with basin rings	(2) Cannot be responsible for inner rings unless excavation cavity	(2) Strength differences probably negligible at basin scales

Source: Adapted from Pike and Spudis (1987)

There is no general agreement on a model for the origin of lunar basin rings and one's preference of mechanism primarily depends upon which model one accepts for the location of the rim of the basin transient cavity.

1.2.4 *Basin ejecta and deposit emplacement*

As large impact craters, basins distribute enormous volumes of *ejecta* widely over the lunar surface. Study of this material has produced intense controversy over its provenance and mechanism of emplacement. Whether or not basin ejecta were sampled by any of the lunar surface missions is still under debate by many scientists. Important insight into the process of basin ejecta emplacement has come from the work of Oberbeck (1975) who extrapolated from the results of laboratory cratering experiments and postulated that the deposits surrounding basins on the Moon incorporate large amounts of locally derived material during the emplacement process. Thus, although the continuous deposits of several basins were directly sampled by the Apollo and Luna missions, there is no assurance that primary ejecta from any basin were sampled as Oberbeck's model predicts the admixture of local *secondary-crater* ejecta with *primary* ejecta within the basin *continuous deposits*. This interpretation has been supported by other workers on photogeologic grounds (Head and Hawke, 1975; Hawke and Head, 1977b), study of the continuous deposits of large terrestrial craters (Hörz *et al.*, 1983), and study of lunar samples from the Apollo 14 and Apollo 16 landing sites (Simonds *et al.*, 1977; Stöffler *et al.*, 1985).

This concept of "local mixing" has been challenged by several workers who believe that primary basin ejecta are present in overwhelming quantities at the Apollo highland landing sites. On the basis of photogeologic arguments, Chao *et al.* (1975) and Wilhelms *et al.* (1980) consider basin primary ejecta to be dominant over large regions of the Moon, and secondary-crater ejecta to be present in abundance only where clearly defined secondary craters can be seen. Thus, some lunar sample investigators prefer the interpretation of Apollo 14 breccias as primary ejecta from the Imbrium basin (Wilshire and Jackson, 1972; Swann *et al.*, 1977). Moreover, theoretical modeling of ejecta behavior during large impacts suggests that significant quantities of primary ejecta may be preserved during the extended times (on the order of minutes) of basin ejecta emplacement on the Moon (Schultz and Mendenhall, 1979; Schultz *et al.*, 1981; Schultz and Gault, 1985).

Levels of shock in basin ejecta potentially can help identify samples at Apollo landing sites that must have been derived great distances from the basin cavity. Material that has been emplaced ballistically must have been ejected from the basin transient cavity at minimum accelerations that may be computed by ballistic theory. Through the use of Hugoniot data from terrestrial analogs, these accelerations can be converted into values of *minimum* shock pressure (Dence, 1976; Austin and Hawke, 1981). Results of these studies suggest that material at the Apollo 14 landing site that was derived from the Imbrium basin must have been shocked to maskelynite grades of shock metamorphism or higher. It has been pointed out, however, that at terrestrial craters, all grades of shock levels may be juxtaposed at any given radial position with respect to the crater (Hörz and Banholzer, 1980; Hörz *et al.*, 1983). Moreover, the discovery of relatively unshocked meteorites derived from the Moon and Mars (e.g., McSween, 1987) indicates that our understanding of the shock history of material during impact events is poor, at best. Results from shock-level studies of lunar samples are therefore not definitive in regard to ultimate provenance.

Impact melts produced by basin-forming events potentially contain a large amount of information on both the mechanics and size of the impact and data on the composition of the lunar crust in the target region. Basin impact melts appear to be concentrated within the inner rings (Head, 1974a; Moore *et al.*, 1974; Scott *et al.*, 1977), and according to terrestrial analogy (Floran and Dence, 1976; Grieve *et al.*, 1977), consist of relatively clast-free melt rocks with a variety of petrographic textures. Photogeologic and petrologic evidence indicates that all melt is not retained solely within the basin, but may be ejected with the clastic material to form melt flows and ponds (Moore *et al.*, 1974; Schultz and Mendenhall, 1979; Spudis and Ryder, 1981).

Studies of impact melts from large terrestrial craters suggest that an impact will homogenize widely diverse chemical compositions (Grieve *et al.*, 1977; Phinney *et al.*, 1978). This also appears to be true on the Moon (Simonds, 1975; Ryder and Wood, 1977; James *et al.*, 1978), but the geological context of lunar samples is poorly constrained and variations in chemical composition of impact melt from a single event cannot be ruled out (Spudis and Ryder, 1981; Spudis, 1984). Problems may also arise through too strict a dependence on terrestrial analogs for the interpretation of lunar samples because of differences in the target rocks and the presence on Earth of volatiles in the target, which can help dissipate shock heat (Kieffer and Simonds, 1980), thus altering the thermal history of melt sheets.

The distribution of basin ejecta with respect to the crater has been the topic of many studies. Attempts to model the radial thickness decay of basin ejecta have produced widely varying estimates (Short and Forman, 1972; McGetchin *et al.*, 1973; Pike, 1974). Recently, emphasis has been placed on the extremely irregular areal distribution of basin ejecta deposits (e.g., Wilhelms, 1987). Local accumulations of primary ejecta may occur and fill pre-existing craters (Moore *et al.*, 1974; Wilhelms *et al.*, 1980) so the generalized models of ejecta thickness decay should be used only with caution.

Although some hypotheses for the origin of basin characteristics appear to be more likely to be correct than do others, each idea is subject to the constraints provided by the lunar data. Integration and synthesis of many types of data may at least partly resolve some of these problems.

1.3 The approach of this book

The study of the geology of lunar basins has produced widely divergent ideas. Most of these concepts have resulted from analysis by some chosen methodology, with little interaction with the results from other disciplines. In this book, I analyze and integrate a variety of lunar data to produce a geological model for lunar basin formation and evolution consistent with the constraints derived from photogeology, remote-sensing, lunar sample studies, study of terrestrial impact features, and analysis of lunar geophysical data.

1.3.1 Photogeological evidence

Although geological maps have been made of the entire lunar surface (Wilhelms and McCauley, 1971; Wilhelms and El-Baz, 1977; Stuart-Alexander, 1978; Scott *et al.*, 1977; Lucchitta, 1978; Wilhelms *et al.*, 1979), further photogeological analysis can contribute to the interpretation of the Moon. Pre-basin topography and structure strongly affect the distribution of basin ring structures and ejecta facies. For example, previous work (Spudis and Head, 1977; Head,1977a; Schultz and Spudis, 1978; Whitford-Stark, 1981a) has demonstrated the important effects of the pre-existing Serenitatis and Insularum basins on the peculiar asymmetries of Imbrium basin morphologies. These pre-basin landforms accentuated basin irregularities and produced departures from the orderly deposition of ejecta facies. More importantly, these features may help delineate the location of the basin transient cavity; if pre-basin structure can be mapped within the present topographic rim, this latter feature cannot represent the transient cavity of the basin.

For all multi-ring basins, the term "circular" is actually only a crude approximation. Basin morphologies are extremely complex and the many departures from a circular plan provide important clues to the processes involved in ring formation as well as implications for regional geological structures. Some basins, such as Imbrium, display morphologic asymmetries that provide information on the location of the transient cavity and the relative timing of movements during the modification stage. Geological mapping of ejecta facies and asymmetries reveal radial and azimuthal variations in the energy levels involved in different ejecta depositional environments.

Almost all studies of lunar basins make the implicit assumption that the initial morphology of the basin (before volcanic flooding) resembled the relatively unflooded Orientale basin. This assumption has been questioned (Schultz, 1979; Spudis *et al.*, 1984a) on the grounds that thermal conditions in the early Moon were probably different than those prevailing at the time of the Orientale impact. The very high heat flow rates early in lunar history lower the mean viscosity of the crust, thus providing a mechanism whereby basin topography and morphology can be easily modified prior to the episodes of mare flooding.

1.3.2 Geology and petrology of Apollo and Luna landing sites

One of the most important sources of data for lunar basin geology comes from the basin-related material obtained by Apollo and Luna sample return missions. There is general agreement that basin impact melt was sampled at the Apollo 17 site (the poikilitic melt "sheet"; Winzer *et al.*, 1977) and possibly at the Apollo 15 site (the "black and white rocks"; Ryder and Wood, 1977; Ryder and Spudis, 1987). Basin impact melt rocks may be present at the Apollo 16 (from Imbrium and Nectaris), Luna 20 (Crisium), and Apollo 14 (Imbrium) landing sites as well. The data base from lunar samples was used to generalize about melt petrogenesis in large impacts, to assess the effects of regional geochemical variations in the highlands on the composition of various basin melts, and to infer the vertical petrologic structure of the crust (Dence *et al.*, 1976; Ryder and Wood, 1977; Spudis and Davis, 1986). At some sites (e.g., Apollo 16), many melt compositions are present (Ryder, 1981a) and this relation may indicate the existence of several melt sheets; some are probably related to basins (Spudis, 1984). The difficulty is to distinguish impact melts produced by small, local craters from those that are basin-related.

In addition to the large collection of impact melts, clastic ejecta from basins may have been collected at the Apollo 14, 15, 16 and Luna 20 sites. Both continuous and discontinuous deposits from basins have been directly sampled and their characteristics provide information on the questions of local versus primary ejecta contained within basin deposits, if likely candidates can be identified. Compositional variations in basin ejecta from the various landing sites on the Moon also provide ground truth for the existence of regional geochemical provinces (Hawke and Spudis, 1980; Warren and Taylor, 1981). If such provinces can be consistently recognized, we may be able to establish the provenance of basin deposits over large regions of the surface of the Moon.

1.3.3 Remote sensing of the chemistry and mineralogy of basin ejecta

The Apollo 15 and 16 orbiting spacecraft carried instruments for the remote measurement of the chemical composition of the surface of the Moon (Adler and Trombka, 1977). Concentrations of Al, Si, and Mg are available for about 12 percent of the lunar surface whereas concentration data for Th, Fe, and Ti cover about 20 percent of the Moon. These data are particularly useful for the study of basins in that they provide regional information on the composition of basin deposits. Moreover, regional variations and trends in the orbital data indicate the extent of geochemical provinces and the relative importance of basin deposits to the overall chemistry of the highland crust. The orbital geochemical data used here are the reductions presented by Bielefeld *et al.* (1976), Adler and Trombka (1977), La Jolla Consortium (1977), Metzger *et al.* (1977), and Davis (1980).

One of the most useful techniques for interpreting orbital geochemical data is the mixing model. In this method, the measured chemical composition of highland regions is interpreted in terms of hypothetical proportions of rock types through the use of a least-squares analysis (Bryan *et al.*, 1969). A mixing model does not indicate what rock types are present; it is merely a tool by which chemical data can be visualized in geological terms. The choice of petrologic end members for mixing models is a longstanding problem; most lunar highland samples display multiple episodes of brecciation, remelting, and chemical mixing. For many years, the search for the primary rock types on the Moon was hampered by this impact metamorphic overprint. I have chosen two groups of end members to model the composition of lunar basin deposits.

The highland compositions described by Taylor (1975) and Ridley (1976) were

Table 1.3 *End member compositions for lunar geochemical mixing models*

End Member	Type	Abbrev.	Ti (wt. %)	Al (wt. %)	Fe (wt. %)	Mg (wt. %)	Th (ppm)
ferroan anorthosite	P	FAN	0.012	18.8	0.18	0.06	0.01
gabbroic anorthosite	M	GABAN	0.21	16.4	2.7	2.0	0.23
anorthositic gabbro	M	ANGAB	0.22	13.7	4.1	4.4	0.73
norite	P	NOR	0.10	11.1	3.9	7.1	1.7
troctolite	P	TROC	0.03	10.0	3.8	11.5	0.9
low-K Fra Mauro basalt	M	LKFM	0.75	10.0	7.6	6.6	5.3
KREEP basalt	P	KREEP	1.32	7.8	8.3	4.9	10.5
Apollo 11 mare basalt	P	11MB	6.3	5.5	14.4	4.2	1.0
Apollo 12 mare basalt	P	12MB	1.7	4.5	16.4	7.0	0.9
Apollo 15 mare basalt	P	15MB	0.9	4.9	14.5	5.7	0.6
Apollo 17 mare basalt	P	17MB	7.3	4.9	16.8	4.9	0.5

Note: P = pristine rock type; M = mixed rock type.
Source: Data for pristine rock compositions from Ryder and Norman (1978a, 1978b) and Taylor (1975). Data for mixed rock type compositions from Taylor (1975) and Ridley (1976).

used to generate the first set of mixing models (Table 1.3). These compositions are those of large hand samples of rock types commonly found in the lunar highlands. In particular, the compositions "anorthositic gabbro" and "low-K Fra Mauro basalt" appear to be impact mixtures of other primary rock types (Keil *et al.*, 1975; Reid *et al.*, 1977). However, these end members provide an estimate of the relative abundance of rock and breccia types that would be expected from samples taken at sites unvisited by Apollo or Luna spacecraft (Hawke *et al.*, 1980).

The second set of mixing models uses end member compositions derived from study of the large collection of "pristine" highlands rocks (Table 1.3), studied in detail by Warren and Wasson (1977) and summarized by Norman and Ryder (1979) and Warren (1985). Pristine rocks on the Moon are those derived solely from internally generated magmas and are not remelted or mixed with other rock types by impact events, although they frequently display the effects of mechanical brecciation. These samples are presumed to represent the primary "building blocks" of the igneous basement of the lunar crust. Mixing models performed with these end members allow us to estimate the petrologic makeup of highland regions and the possible nature of the crustal basement at basin target sites. Thus, the two end member selections are complementary, both providing different ways of visualizing the orbital geochemical data for direct study of basin ejecta deposits (Spudis and Hawke, 1981).

Other clues to the composition of the lunar surface come from Earth-based spectrophotometry (McCord *et al.*, 1971, 1981; Pieters, 1986), a technique that permits the identification of glass and mineral species in lunar surface materials. The essential mineralogy of most lunar highlands rocks is quite simple: plagioclase feldspar dominates and the major mafic phases are olivine and clinopyroxene or orthopyroxene (Taylor, 1982; Warren, 1985). Although spectra cannot yet quantify the proportions of mafic minerals present (necessary to establish the mode of a rock and thus, its petrologic classification), they can identify likely candidate rock types because lunar highlands rocks possess this simple mineralogy. Spectral data have been used extensively in the study of basin deposits (e.g., Spudis *et al.*, 1984b, 1988b) and, when combined with data from orbital chemistry and Apollo and Luna site geology, can provide insight into regional variations in these deposit compositions.

1.3.4 Geophysical data

Several types of lunar geophysical data provide constraints on basin impact mechanics and regional geology. An important data source is the global map of crustal thickness derived from orbital gravity data by Bills and Ferrari (1976). This map estimates regional variations in crustal thickness around each lunar basin studied and was calibrated by crustal thickness values derived from the Apollo seismic stations (Goins *et al.*, 1979). Estimates of crustal thickness variations obtained independently from isostatic modeling of lunar highland regions through the use of orbital gamma-ray Fe values (Haines and Metzger, 1980) suggest that this map is largely accurate. This map constrains the probable petrology of the basin target as well as directly indicating the depth to the lunar mantle in a given target region, thereby allowing an estimate of what depth would have to be excavated to eject samples from the ultramafic mantle of the Moon.

Heat flow rates obtained at the Apollo 15 and 17 landing sites provide upper limits for the abundances of radioactive U and Th present in the Moon (Langseth *et al.*, 1976). Because the geochemical behavior of these elements suggests they are concentrated in the lunar crust, their abundances permit an estimate of the bulk composition of the crust in the Imbrium and Serenitatis basin target regions (Ryder and Wood, 1977). These estimates may then serve as references when studying the results from the orbital mixing models.

Magnetic data were collected both from the Apollo landing sites and from lunar orbit. It has been proposed that basin ejecta blankets acquired a uniform remanent field during emplacement, induced by cooling through the Curie point in the presence of a global, dynamo-derived magnetic field (Runcorn, 1982), but I doubt this interpretation. Basin deposits such as the Fra Mauro Formation of the Imbrium basin are heterogeneously magnetized and display random local field strengths and orientations (Hood, 1980). Most of the various magnetic anomalies seen on the Moon appear to be related to post-mare impact processes (Hood, 1980; Schultz and Srnka, 1980) and thus, are not directly applicable to basin problems.

1.3.5 Synthesis

The different lunar data described above were all used to derive geological models for five lunar basins: Orientale, Nectaris, Crisium, Serenitatis, and Imbrium (Figure 1.1). These basins were selected because of maximum data coverage from all the sources discussed above; Orientale was included, despite the absence of a sample-return mission, because it is well preserved and historically has been the prototype lunar multi-ring basin. These data constrain the location and relative dimensions of the basin transient cavity and these in turn constrain the probable origin for basin ring structures. The petrologic and geochemical properties of basin deposits are inferred primarily from the geology and petrology of Apollo and Luna highland sites and from the remote-sensing data. The resulting synthesis of lunar basin geology leads to a model for the mechanics involved in basin formation and geological evolution as well as the relation of these features to the history of the lunar highlands crust. The model developed for lunar basins has direct application to the interpretation of multi-ring basins on the other planets and provides insights into their early history.

2

From crater to basin

Multi-ring basins are features produced by the collision of solid bodies with the planets, so the basin problem is a subset of the more general problem of impact cratering, a vast field of study. This chapter briefly describes the impact process from theoretical considerations, from the evidence of some well studied terrestrial impact craters, and from the observed morphology of impact craters on the Moon and their systematic changes with increasing crater size.

2.1 The cratering process

2.1.1 Impact mechanics

Our understanding of what happens when a solid body hits a planetary surface at high speeds has increased greatly over the past 25 years. The study of the physical processes occurring during impact events is called *impact mechanics*. Although the details of this complex process are not understood, laboratory experiments, explosion craters, natural impact craters, and computer simulations have given us a general outline of the main stages that characterize the formation of an impact crater.

Solid bodies collide with planetary surfaces at very high speeds; such impact speeds are in the range called *hypervelocity*. Encounter velocities can vary from lunar escape velocity (about 2.5 km/s) at a minimum, up to many tens of kilometers per second (on the basis of velocities of bodies in heliocentric orbits). On the Moon, the mean impact velocity is about 20 km/s (Shoemaker, 1977). At the moment of contact between an impactor and a planet, the kinetic energy of the impacting body is transferred to the planetary surface target. A *shock wave* propagates into the target and projectile, resulting in intensive compression of both objects. In hypervelocity impacts, the quantities of energy produced greatly exceed the heat of vaporization for geological materials. Thus, both the projectile and some fraction of the planetary surface are vaporized. Passage of the shock wave through the planetary target and the projectile is followed by a decompression (rarefaction) wave. As these decompression waves from the compressed target encounter the free surface of the planet, material begins to be ejected from the target, forming a crater.

At this point, most of the kinetic energy contained by the speeding projectile has been transferred to the planetary surface target. This transfer creates a cratering *flow field* (Figure 2.1). Within this flow field, materials that make up the target are in motion and follow spiral paths away from a central point (Stöffler *et al.*, 1975; Maxwell, 1977; Croft, 1980). Material directly beneath the impact point is driven downward while target materials slightly away from this area are driven both downward and outward (Figure 2.1). Maxwell (1977) and Croft (1980) have

Figure 2.1 Schematic diagram of the excavation and modification phases in the formation of a simple, bowl-shaped crater. In (a) a cratering flow field has been established, with target material flowing outward from a single point in logarithmic spirals (Maxwell, 1977; Croft, 1980). Only material above the hinge streamline is ejected from the crater (shaded); this area constitutes the excavation cavity. The transient cavity is the bowl-shaped feature within which material is in motion during cratering flow. Part (b) shows the relation between the apparent and true craters after the crater has formed; note that neither of these two features corresponds to the geometric constructs present during cratering flow. Diagram by R.A.F. Grieve from Heiken *et al.* (1991).

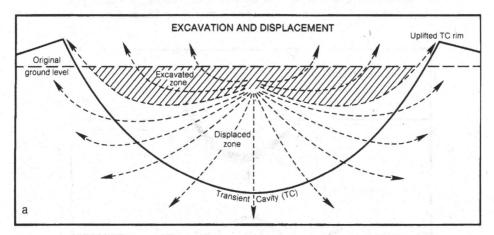

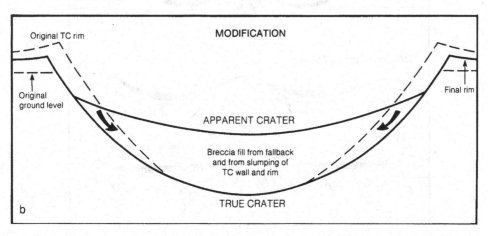

adapted a simple analytical scheme, called the "Z-model", to describe the cratering flow field. The Z-model predicts that material in the target follows flow streamlines that take the form of logarithmic spirals, originating from a single point. All material within the crater is in motion, flowing away from the center (Figure 2.1); material below a certain critical flow line (the *hinge streamline*) remains within the crater whereas target material above the hinge streamline is excavated from the crater.

Because the intensity of the shock wave decreases radially away from the point of impact, the target materials experience varied degrees of shock damage during an impact event. The process of modification of geological materials by shock waves is called *shock metamorphism* and the grades of shock damage to the target rocks decrease as the distance from the point of impact increases. Material closest

Figure 2.2 The modification stage in the formation of a complex crater (i.e., one that displays terraces, flat floors, and interior peaks). After excavation (a), rebound of central zone beneath crater thrusts upward, forming central uplift (b). The uplift subsequently collapses (c), in conjunction with collapse of rim. In final form (d), crater consists of slumped wall material and central peak and/or inner ring; impact melt sheet lines crater floor. Diagram by R.A.F. Grieve from Heiken *et al.* (1991).

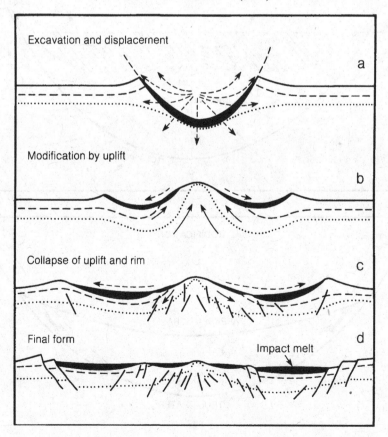

to the impact point experiences the highest peak pressures and may be completely vaporized. Slightly farther from the impact point, the peak pressure is still high enough to completely (not partially) melt the target; the resulting material is called *impact melt* (Figure 2.2) and lines the floor of the growing cavity. In terrestrial impact craters (discussed below), impact melts tend to be chemically homogeneous averages of the target rocks. Continuing away from the impact point, target rocks are progressively less shocked, although many retain signatures of the shock metamorphism, as shown by damaged mineral crystals. This damage includes surface dislocations, formation of diaplectic glass, and partial melting along grain boundaries in the shocked target rocks.

Some confusion has resulted from the cratering literature because people use various terms, and in more than one way, to describe an impact crater during the excavation phase. The crater *transient cavity* is, as the name implies, an ephemeral feature within which all materials are in motion as a result of the cratering flow. For simple, bowl-shaped craters, the transient cavity has a depth–diameter ratio of about 1:3; whether this relation remains the same for larger craters is a subject of controversy and will be discussed in several places in this book. The crater *excavation cavity* is a feature identical to the crater transient cavity in diameter, but quite different in shape and depth (Figure 2.1). It is from this zone in the target that material is actually removed to form the crater, ejected along ballistic paths through space, and ultimately deposited around the final crater as an *ejecta blanket*. Although the shape and dimensions of the crater excavation cavity vary according to factors such as projectile density and impact velocity, empirical evidence from experiments (Stöffler *et al.* 1975), analytical models (Croft, 1980), and the geology of terrestrial simple craters (Grieve, 1980a, 1987; Grieve and Garvin, 1984) suggests that the maximum depth of excavation is 0.08 to 0.12 times the diameter of the excavation cavity.

As the transient cavity grows, material is ejected continuously from the crater, generally at angles around 45 degrees (Gault *et al.*, 1968). The ejecta take the form of an inverted cone of material in motion radially outward from the point of impact; materials thrown out of the crater earliest, and at the highest velocities, travel farthest from the point of impact. Excavation of target material continues until the planet's gravity overcomes the force of ejection induced by the decompression wave. At this point, the *excavation phase* of crater formation is completed and is immediately followed by the *modification phase* (Figure 2.2). In the ejecta cone, material is deposited on the surface of the planet; the last material ejected is deposited first, nearest the rim. Within the crater itself, floor materials rise to compensate for the excavated mass; in small, bowl-shaped craters, this rebound is minor but in larger craters, it can form central mounds or peaks (Figures 2.2, 2.3).

The crater modification stage is less well understood than the excavation phase, particularly as the size of the cratering event increases. Slumping of material from the crater walls into the floor occurs, the crater *melt sheet* lines the floor and begins to cool, and the ejecta blanket is deposited. The resulting crater (Figure 2.1) consists of two features: the *apparent crater* is the actual depression in the pre-crater planetary surface that is visible; it does not include the raised rim material. The *true crater* (also referred to as the "strength crater"; Croft, 1980) encompasses the

Figure 2.3 Comparative cross sections showing the relations between geological units at terrestrial simple (a) and complex (b) craters. Diagram by R.A.F. Grieve from Heiken *et al.* (1991).

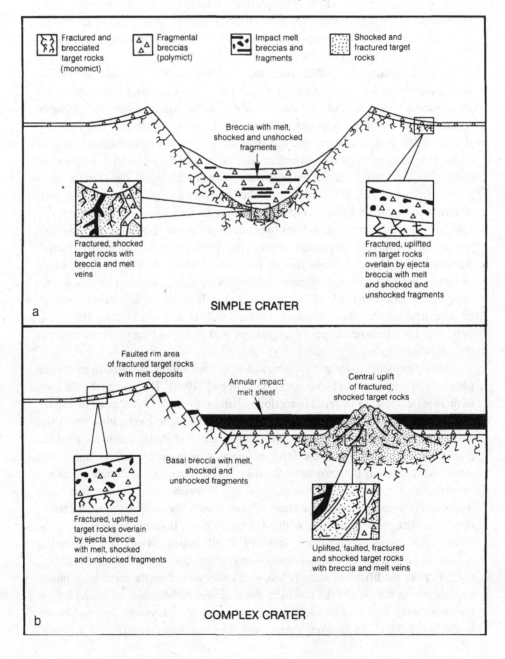

apparent crater, but also includes brecciated target material, most of which was in motion during cratering flow, and structurally displaced material associated with the rim and floor units (Figure 2.1b). Thus, the crater observed in planetary photographs has little correspondence with any of the transient crater phases present during the excavation stage of crater formation. The crater has taken only minutes to form, a rare example of an "instant" geological process; however, the time of crater formation increases as the size of the event increases. For basin-sized impacts, total formation time may take from tens of minutes to hours (Gault, 1974).

2.1.2 Impact craters: the terrestrial example

Under the impetus of lunar exploration in the 1960's, the search for and study of terrestrial impact craters greatly increased and a gradual recognition developed of the importance of impact cratering in the early geological history of the Earth (Shoemaker, 1977; Baldwin, 1978). Few impact craters are preserved on the Earth because the dynamic processes of volcanism, erosion, sedimentation, and plate tectonics have erased its most ancient surfaces. However, over 120 impact craters on the Earth are now known (Grieve and Robertson, 1979; Masaitis *et al.*, 1980; Basilevsky *et al.*, 1983; Grieve, 1987) and study of these features has greatly improved our knowledge of the cratering process.

Three terrestrial impact craters are relatively well preserved and display features that have proven particularly useful in understanding the geology of impact craters and basins on the Moon. The Manicouagan crater, Quebec, has an impact melt sheet that has been used as the standard model for understanding the impact melts of lunar basins (Simonds *et al.*, 1976). The Ries crater, Germany, retains a fairly well preserved continuous ejecta blanket that illuminates details of the ballistic sedimentation process (Pohl *et al.*, 1977; Hörz *et al.*, 1983). The Popigay crater, Siberia, displays several features applicable to the interpretation of lunar craters and basins including a melt sheet, ejected melt masses, multiple rings, and continuous ejecta (Masaitis *et al.*, 1976, 1980). The geology of these craters will be reviewed briefly, with emphasis on those features that appear to be relevant to the geological interpretation of lunar basins.

The Manicouagan crater (Figure 2.4) is located on the Canadian shield in northern Quebec. The structure displays three circular features; a central plateau region about 25 km in diameter, a circular ring-trough filled by a lake about 65 km in diameter, and a faint outer ring fault (not shown here) about 150 km in diameter (Floran and Dence, 1976). The impact occurred in Precambrian crystalline rocks of widely varying chemistry and petrology, ranging in composition from gabbro to granite (Phinney *et al.*, 1978).

The original size of the Manicouagan crater is subject to debate. A consortium of workers who studied the structure from a multidisciplinary approach favored a relatively small transient cavity, about 30 km diameter (Phinney *et al.*, 1978). The model of Orphal and Schultz (1978) suggests that the original crater was much larger (more than 55 km) and the present morphology of the crater is due to internal modification and floor uplift, similar to lunar floor-fractured craters (Pike, 1971; Schultz, 1976a). A re-analysis of the dimensions of Manicouagan by Grieve and Head (1983) suggests that it is structurally equivalent to the lunar complex crater Copernicus and that no evidence exists for internal modification. If the interpretation of Phinney *et al.* (1978) is correct, Manicouagan is a multi-ring structure; if the latter interpretations are correct, it is probably comparable to lunar complex craters of the central-peak type (Grieve and Head, 1983).

The central crater region is dominated by a planar sheet of fine-grained impact melt. This melt sheet is remarkable for its chemical homogeneity and petrographic heterogeneity (Simonds *et al.*, 1976; Floran *et al.*, 1978). The melt has varied clast abundances, but typically displays less than 15 percent clasts by volume (Floran *et*

Figure 2.4 The Manicouagan impact crater, Quebec. The circular black pattern, the Manicouagan reservoir, outlines the original crater rim (65 km diameter). The crater impact melt sheet is confined inside this bounding structural trough and overlies shocked, Precambrian crystalline rocks. NASA Landsat photograph.

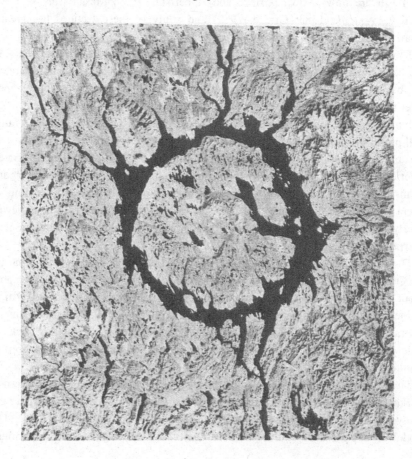

al., 1978), which corresponds to lunar impact melts characterized as clast-poor. The chemical homogeneity of the melt sheet is even more remarkable, considering that the impact target consisted of such diverse lithologies as granite, granitic gneisses, anorthosite and metagabbro (Grieve and Floran, 1978).

The Manicouagan melt sheet has several features similar to lunar impact melt rocks. Petrographic texture is strongly influenced by clast content (Simonds *et al.*, 1978), just as is observed in lunar impact melts (Simonds, 1975). The chemical homogeneity of Manicouagan impact melts is analogous to that observed in the Apollo 17 poikilitic melt rocks (Simonds, 1975; Spudis and Ryder, 1981). Clastic ejecta have been removed by erosion, but concentric structural elements appear to be well preserved (Dence, 1977b). Particularly interesting is the structural ring trough surrounding the main melt sheet, suggesting that post-impact structural adjustment at Manicouagan may have continued for some time after the initial post-impact modification stage (Orphal and Schultz, 1978).

The Ries crater of southern Bavaria is one of the earliest recognized terrestrial impact craters (Shoemaker and Chao, 1961). This feature is about 22 km in diameter and displays a circular ring within the main crater rim (Figure 2.5; Pohl *et al.*, 1977). The impact target consisted of a thin layer of sedimentary rocks overlying a crystalline basement complex. The continuous ejecta from the Ries crater, called the "Bunte breccia", extend for about a crater diameter from the rim and are discon-

Figure 2.5 Geological sketch map of the Ries impact crater, Bavaria. The crater rim (dashed) contains suevite deposits (ejected melt bombs and highly shocked debris); continuous deposit (Bunte breccia) is largely clastic, occurs outside this ring structure, and is exposed mostly on the southern side of the crater; suevite locally overlies the Bunte breccia. From Taylor (1982), after Gall *et al.* (1977).

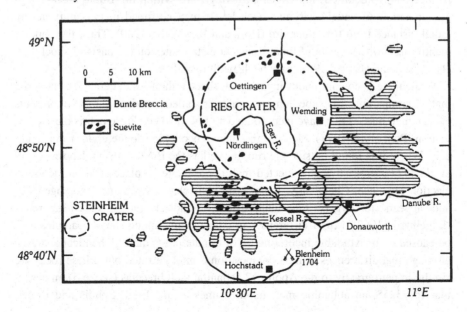

tinuous to absent in many sectors because of post-impact erosion (Schmidt-Kaler *et al.*, 1970). This material has been intensely studied by lunar geologists as a model for the continuous deposits surrounding lunar basins (Chao, 1974; Oberbeck, 1975; Hörz *et al.*, 1983). Also present in abundance are aerodynamically shaped melt bombs that contain shocked crystalline clastic debris. This material, called "suevite", appears to represent the impact melt produced by the Ries event (Engelhardt, 1967; Kieffer and Simonds, 1980), but it never formed a continuous melt "sheet".

Petrography of the Bunte breccia suggests that the large amounts of local material incorporated into the deposits were mixed with primary ejecta during the emplacement process (Oberbeck, 1975; Hörz and Banholzer, 1980; Hörz *et al.*, 1983). The quantities of substrate incorporation have been debated (e.g., Chao, 1974), but the available data suggest that such incorporation did occur. A relevant question is to what degree the processes of ballistic sedimentation (Oberbeck, 1975) are responsible for the incorporation of local materials.

The evidence for ground flow of the Bunte breccia deposits is compelling (Chao, 1974; Pohl *et al.*, 1977); this ground flow may have been similar to the flow of ejecta ramparts seen around martian craters, which have been attributed to the entrainment of volatiles within the crater ejecta (e.g., Carr *et al.*, 1977). Because the impact target of the Ries crater probably contained large quantities of ground water, a similar mechanism is envisioned for the post-depositional ground flow of the Bunte breccia (Kieffer and Simonds, 1980). Thus, although not directly comparable to lunar basin ejecta, study of the Ries Bunte breccia suggests that the hypothesis of local mixing in continuous deposits is valid; whether the absolute quantities of locally derived material seen in the Bunte breccia can be extrapolated to the Moon is uncertain. In addition, individual clast fragments within the Bunte breccia suggest materials of widely varying shock levels may be found in proximity at any radial distance from the crater rim (Hörz and Banholzer, 1980). Thus, the hope of identifying primary ejecta of lunar basins at distant sites on the basis of shock levels (Austin and Hawke, 1981) may be difficult to realize.

While there is no continuous impact melt sheet at the Ries crater, a Ries material that is comparable to some lunar samples is called suevite. Suevite represents ejected melt masses that have incorporated highly shocked clastic debris during formation (Pohl *et al.*, 1977). Much of the suevite occurs inside the crater rim (Figure 2.5). Where it is found outside the rim, it overlies the Bunte breccia deposits (Pohl *et al.*, 1977), suggesting that ejected melt deposits are emplaced late in the crater ejection sequence. Chemically, the suevite melt phase is relatively homogeneous and appears to largely represent fusion of the crystalline basement rocks (Engelhardt, 1967). Minor differences in composition among suevite samples may be caused by unavoidable incorporation of minute clasts during chemical analysis, although real differences in melt composition cannot be ruled out. These rocks are similar in petrography to the Apollo 17 aphanitic melt breccias (sec. 6.5), in particular to 73255, an aphanitic melt bomb (James *et al.*, 1978; Spudis and Ryder,

1981). The Ries results suggest that ejected melt masses from large cratering events may be as homogeneous as the main crater melt sheet.

The Popigay crater in Siberia is about 100 km in diameter (Figure 2.6) and displays at least two rings interior to the crater rim of 45 and 70 km diameter and possibly an exterior ring 140 km in diameter (Masaitis *et al.*, 1976, 1980; Pike, 1985). The target varied in lithology vertically and laterally and consisted of thin Mesozoic and Paleozoic sediments overlying Archean gneiss. The crater is recognized as a broad, topographic depression and contains remnants of an impact melt sheet, abundant fallback breccia, a thick lens of suevite, and patches of continuous ejecta that make up the crater rim (Figure 2.6; Masaitis *et al.*, 1976). All interior deposits of the crater mantle a series of thrust faults, ring troughs, concentric horsts, and uplifted basement. Evidence from Popigay suggests that basin rings are structurally complex, uplifted from great depths, and that these uplifted basement materials are intercalated with shock-processed materials and fallback ejecta produced during basin excavation.

The impact melts of Popigay crater, called tagamites after the Tagam hills in the eastern part of the crater, display remarkable chemical homogeneity and highly varied clast abundances and petrographic textures. The chemical composition of the tagamites is a mixture of the rock types present mostly in the crystalline basement (Masaitis *et al.*, 1976), suggesting that the domain of melt petrogenesis occurs within the lower portions of the transient cavity. Ejected melt masses and dikes and pods of crater melt all appear to be chemically homogeneous (Masaitis *et al.*, 1976). Melt fragments that occur within the massive suevite deposits are slightly richer in silica and alumina than the tagamites and also show slightly more compositional dispersion. Such properties may indicate a slightly greater admixture of the target sedimentary rocks in the melt sheet.

The rings of the Popigay crater warrant consideration. Although the crater is significantly eroded, being of Tertiary age (39 Ma old; Masaitis *et al.*, 1980), detailed mapping of the structure has revealed several large-scale structural uplifts, whose axes can be reconstructed with fair precision (Figure 2.6a; Masaitis *et al.*, 1980; Pike, 1985). Popigay appears to be a true multi-ring basin, with clear evidence for at least two rings inside the rim of the crater and less definitive evidence for another ring outside the rim (Figure 2.6a). These rings are spaced at a constant interval ($\sqrt{2}\ D$), identical to that exhibited by multi-ring basins on the Moon and other planets (Pike and Spudis, 1987).

Terrestrial craters have a venerable history as analogs for lunar basins, but some caution must be exercised when conclusions are drawn about the cratering process. The Earth is a volatile-rich, dynamic body that is not entirely comparable to the targets for basins on the Moon. Nonetheless, several important geological features of terrestrial craters appear to be similar to phenomena observed on the Moon. Melting by impact is a process whereby rocks of widely diverse composition may be fused into a chemically homogeneous mass. This process appears to be valid for melt that is ejected from the cavity as well as the melt mass that remains within the central crater to form the melt sheet. Continuous ejecta deposits may incorporate large amounts of local material during deposition, but the exact quantity is uncertain. These continuous deposits contain a melange of materials from the crater target, exhibiting varied degrees of shock metamorphism. The post-event style of modification of terrestrial craters appears to depend strongly upon local geological conditions, but long-term structural adjustment apparently can be important in determining the final morphology of large craters. A morphological sequence of simple bowls to multi-ring basins is evident in the terrestrial record and study of

Figure 2.6 Geological sketch map (a) and cross section (b) of the Popigay crater, Siberia. The crater is a multi-ring structure, displaying four rings of 45, 70, 100, and 140 km diameter. As at the Ries crater, suevite is confined largely within the inner basin; these rocks are injected by dikes of fine-grained melt called tagamites. After Masaitis *et al.* (1980).

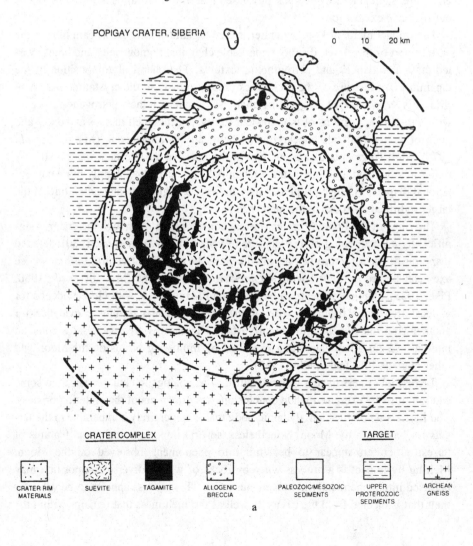

POPIGAY CRATER, SIBERIA

0   10   20 km

CRATER COMPLEX   TARGET

CRATER RIM   SUEVITE   TAGAMITE   ALLOGENIC   PALEOZOIC/MESOZOIC   UPPER   ARCHEAN
MATERIALS   BRECCIA   SEDIMENTS   PROTEROZOIC   GNEISS
                                                                                                 SEDIMENTS

a

individual morphological elements on the Earth may provide a guide to the geological interpretation of such features on the planets. Thus, the data for terrestrial craters provide direct evidence for several processes inferred to be important during impacts on the Moon and other planets.

2.2 The morphology of fresh lunar craters

Impact craters on the planets display uniform changes in morphology with increasing crater size (Pike, 1980a,b,c). On the Moon, study of changes in morphology as a function of size has been concentrated on fresh craters; it is assumed that all lunar craters originally resembled these features and that subsequent degradation is responsible for current morphologic differences. Although this is a reasonable assumption for virtually all lunar craters, it may not be strictly applicable to lunar basins, which formed on a hot, nascent Moon having target properties that were probably quite different from either the current Moon or any terrestrial analog.

All lunar craters less than about 15 km in diameter display a simple bowl-shaped form (Figure 2.7a); this form represents the shape of the *apparent crater* (Figures 2.1, 2.3), a feature different from both original crater transient cavity and the excavation cavity. The central zone of non-excavated debris beneath the origin of cratering flow is driven downward and outward during the excavation phase, resulting in rim uplift (Grieve and Garvin, 1984). An annular zone of textured material surrounds the crater rim; this zone consists of crater ejecta and material of the local surface churned up by the impact of such ejecta. Although for simple craters the incorporation of local material into the crater continuous deposits is very minor, it becomes more important with increasing crater size. Beyond the continuous crater deposits is a zone of coalesced, satellitic craters. These features, called *secondary craters*, are formed by the impact of blocks or clouds of primary ejecta thrown out during excavation of the primary crater. Secondary craters are found around all planetary impact craters and become larger with increasing primary crater size.

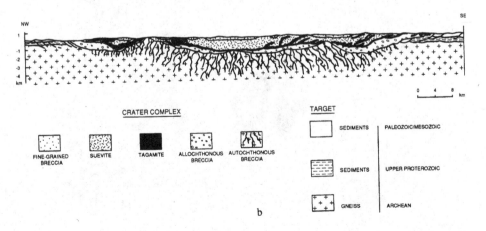

b

Figure 2.7 Changes in crater morphology with increasing crater size. (a) Mösting C, a simple, bowl shaped crater (2.2° N, 8.8° W; 4 km diameter). Craters in the size range below about 15 km display the simple interior of the apparent crater. Textured deposits surrounding the crater form the ejecta blanket; the "v-shaped" texture seen on the ejecta blanket (called herringbone texture) points toward the source crater. Part of LO III-113 M. (b) Dawes, an incipient complex crater (17.2° N, 26.4° E; 18 km diameter). The simple bowl shape has been modified by collapse of the crater walls inward to form debris mounds in the crater interior. Note exposures of mare basalt bedrock in upper crater walls. Part of AS17-2762 (pan). (c) Tycho, a complex crater (43.3° S, 11.2° W; 85 km diameter). Complex craters display flat floors, central peaks, terraced walls, and complex rim morphology. The smooth units on the rim of Tycho are probably impact melt deposits, splashed over the rim during the modification phase of crater formation. Portion of LO V-125 M. (d) Compton, a lunar protobasin (56° N, 105° E; 175 km diameter). Protobasins resemble complex craters in that they have central peaks and terraced walls, but they also display a ring of isolated peaks on the floor between the central peak and floor–wall contact. Fractures on floor suggest some type of floor uplift or cooling of thick melt sheet. Portion of LO V-181 M. (e) Schrödinger, a mature, two-ring basin (76° S, 134° E; 320 km diameter). In this class of crater, no vestige of a central peak remains, but a massif-like inner ring (150 km diameter) is prominent and the crater retains crisp wall terraces. Note floor fractures, one of which displays a low-albedo, volcanic vent. Portion of LO IV 8-M. (f) Korolev, an incipient multi-ring basin (4° S, 158° W; main rim, 440 km diameter). The prominent inner ring (220 km diameter) makes Korolev appear to be a two-ring basin, but subtle massifs and topographic highs between the inner ring and wall and outer fragmentary scarp (arrows) indicate that this feature is a true multi-ring structure (Croft, 1981b; Pike and Spudis, 1987). Portion of LO I-38 M. (g)

Orientale, a mature multi-ring basin (19° S, 95° W; main rim, 930 km diameter). At least four, and possibly six, concentric rings are evident; see Chapter 3 for a detailed discussion of the geology of this basin. LO IV-187 M.

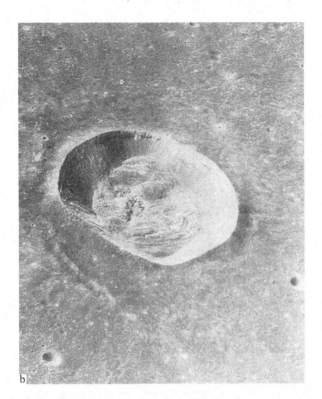

For legend see page 30.

d

e

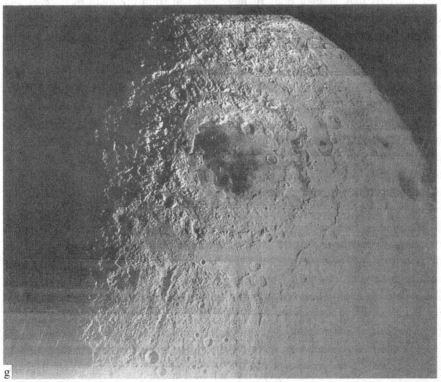

Secondaries can assume a wide variety of complex morphologies, including open and closed clusters, chains, and gouges.

Crater form on the Moon becomes increasingly more complex in the diameter range between 15 and 25 km. This largely results from more extensive modification in the late stages of crater formation; walls of larger craters become less stable and tend to collapse into the crater after excavation ceases, forming debris mounds and slump material on the floor and scalloping the upper walls of the crater (Figure 2.7b). With increasing crater size above about 20 km in diameter, central peaks appear; this class of crater is best illustrated by Tycho, a young 85-km-diameter crater prominent on the lunar near side (Figure 2.7c). Tycho displays a central peak complex, rough textured floor topography, wall terraces, a complex ejecta blanket that includes pools of smooth material, numerous secondary craters, and one of the most prominent ray systems on the Moon. The rough floor materials show evidence of flow; it probably consists largely of impact melt (Shoemaker *et al.*, 1968; Howard and Wilshire, 1975; Schultz, 1976a), the final emplacement of which continued during the crater modification stage. The smooth pools of material evident in the near-rim deposits probably also consist of impact melt (Shoemaker *et al.*, 1968), thrown out during final adjustment of the crater and mantling the rim ejecta units (Howard and Wilshire, 1975; Hawke and Head, 1977a). These observations suggest that basin floors contain impact melt sheets, largely in place, and the near-rim materials may contain pods of melt detached from the main melt sheet and thrown out of the crater towards the end of the crater modification stage.

Central-peak craters on the Moon persist to diameters of 150 to 200 km; part of the reason for the uncertainty of this transition is that *fresh* craters of larger size are rare on the Moon. The larger central-peak craters display floor hummocks that apparently become more prominent with increasing crater size (Hale and Grieve, 1982). Eventually, small rings of peaks are formed; the morphology of these craters suggests they are incipient basins, or *protobasins* (Pike, 1982, 1983). Lunar protobasins (Figure 2.7d) possess some morphologic attributes of basins, in the form of peak rings or ring-like elements, but by all other criteria, are similar to craters, in that they have wall terraces and true central peaks. Pike (1983) recognized that lunar protobasins follow morphometric trends distinct from complex craters and basins (see below), but still possess morphological attributes of both.

The smallest true basins on the Moon are those that display at least two concentric rings; Schrödinger (Figure 2.7e) is the type example. Schrödinger (320 km diameter) strongly resembles a complex crater except for its inner, discontinuous circular peak ring. The peak ring of Schrödinger is not perfectly circular (Figure 2.7e); this irregularity illustrates some of the problems in recognizing and determining the diameters of basin rings, many of which are very degraded and barely visible. The floor of Schrödinger displays fractures and volcanic vents, suggesting that most of the basins filled with mare basalt originally may have possessed similar appearing units. Two prominent radial gouges traverse the exterior ejecta deposits of Schrödinger (Figure 2.7e); these features are similar to radial texture observed around the larger basins. Such radial texture has long been attributed to radial fractures (e.g., Scott, 1972a); an alternative model holds that they are caused largely during secondary cratering by low-angle ejecta thrown out of the crater late in the excavation process (e.g., Head, 1976a).

The transition diameter from two-ring to multi-ring basins is obscure. This uncertainty is partly an artifact of the small sample of basins in this size range on the Moon and their relatively poor preservation state. However, new mapping of all lunar basins has shown that many basins in the 400 to 600 km diameter range, previously thought to be two-ring basins, are in fact, true multi-ring structures (Pike and Spudis, 1987); a prime example is Korolev (Figure 2.7f). The floor of Korolev shows small peaks between the inner prominent ring and the terraced wall; these peaks tend to be found at constant radial distances from the basin center and probably constitute the "missing" ring of the basin. Additionally, concentric scarps outside the main rim (Figure 2.7f) suggest that large, exterior rings are incipient. Although not well developed, the presence of these features indicates that Korolev is a true multi-ring basin.

The famous Orientale basin (Figure 2.7g) displays all of the classic attributes of lunar multi-ring basins, including at least four (and possibly as many as six) concentric ring structures. The geology of this feature is dealt with at length in the next chapter.

2.3 Size-dependent morphologic thresholds: crater to basin

As described in the preceding section, the morphology of impact craters is strongly dependent upon crater size. The diameters at which these distinct morphological transitions occur is a function of planetary surface gravity and target composition (Pike, 1980b; 1980c; 1988); gravity appears to be the most important variable (Gault *et al.*, 1975), although the effect of target properties is still not fully understood and may be of great importance (Cintala *et al.*, 1977). The simple-to-complex crater transition is the one best determined for all the planets, mainly because the statistical sample of such features is the largest. For the purpose of basin study, however, the most important transitions are those for complex craters to protobasins (cf. Figures 2.7 c,d), protobasins to two-ring basins (cf. Figures 2.7 d,e), and two-ring to multi-ring basins (cf. Figures 2.7 e,f).

On the Moon, the size range for complex craters and protobasins overlaps (Figure 2.8) but generally falls between 140 and 200 km (Pike, 1983). This transition may even begin at much lower diameters; Hale and Grieve (1982) have noted that the larger complex craters (>80 km diameter) show increased floor roughening, possibly representing the incipient growth of inner ring morphology. At least three of the Moon's most prominent protobasins occur near Schrödinger (Figure 2.7e), suggesting that some contribution from target conditions may be important in the formation of these features. The transition between protobasins and two-ring basins is obscure and overlaps, but certainly it occurs between 250 and 300 km (Figure

2.8). This transition is marked by a gradual disappearance of the remnant central peak, more mature morphologic expression of the inner peak ring, and a distinct movement of diameter ratios to the well-defined 2:1 ratio that describes the basin rim/inner ring diameters for two-ring basins on all the terrestrial planets (Wood and Head, 1976; Wood, 1980; Pike, 1983; Pike and Spudis, 1987).

The transition from two-ring to multi-ring basins is the most poorly known of all the transitions of planetary crater morphology. Early basin studies concentrated on the circular maria of the lunar near side, the largest members of the multi-ring class (Baldwin, 1949; Hartmann and Kuiper, 1962). After the Lunar Orbiter spacecraft returned photographs of most of the Moon, including the far side, systematic study

Figure 2.8 Size-dependent morphologic transitions for lunar complex craters (shaded), protobasins (circle/triangle), two-ring basins (dots), and multi-ring basins (square and crosses). The horizontal axis (D_m) is the crater main rim diameter (in km) and the vertical axis (D_n) is the diameter of the central peak complex in complex craters, peak complex plus peak ring in protobasins, inner peak ring for two-ring basins, and all mapped rings for multi-ring basins. After Pike (1983) and Pike and Spudis (1987).

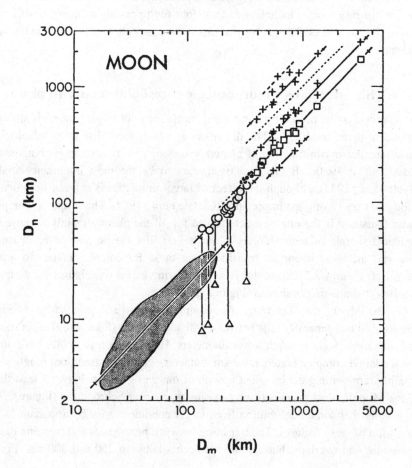

recognized a variety of basins in the 400 to 600 km diameter range (Stuart-Alexander and Howard, 1970; Hartmann and Wood, 1971; Howard *et al.*, 1974). For many years, these features were considered to be two-ring basins but detailed and systematic mapping by Pike and Spudis (1987) showed that some of these basins displayed incipient additional rings, generally appearing between the inner peak ring and the main rim or expressed as large, discontinuous external scarps (e.g., Korolev; Figure 2.7f). Thus, we can now estimate that the transition between two-ring and multi-ring basins on the Moon occurs between 400 and 600 km diameter.

The spacing between rings of multi-ring basins has been studied for many years as it has been thought that this spacing imposes constraints on models of ring formation. It was first noted by Hartmann and Kuiper (1962) that adjacent rings of lunar basins were spaced at constant intervals of either 1.4 D or 2.0 D (i.e., each ring has a diameter about 1.4 or 2 times the diameter of the ring that lies directly inside of it). Fielder (1963) noted that these values were multiples of $\sqrt{2}$ and applied this "rule" to the rings of Orientale (Figure 2.7g) and several other near side basins. These early observations were systematized by Hartmann and Wood (1971), who claimed that the $\sqrt{2}$ spacing "rule" held for *all* lunar multi-ring basins, including a newly described population of basins on the far side of the Moon. The most complete study of basin ring spacing to date (Pike and Spudis, 1987) critically examined this generalization for the Moon, Mercury, and Mars, and found that the $\sqrt{2}\ D$ relation for the spacing of basin rings holds for all three planets. As the spacing of basin rings may be an important constraint for models of ring formation, this topic will be discussed in Chapters 8 and 9.

2.4 Inventory of lunar multi-ring basins

There are more than 40 basins on the Moon (Tables 2.1, 2.2; Figures 2.9, 2.10). Basins are distributed randomly over the lunar surface (Figures 2.9, 2.10; Stuart-Alexander and Howard, 1970). Moreover, their size-frequency distribution forms a continuum with the size–frequency relation observed for large lunar craters. Thus, it is well established that these features are entirely of impact origin (for a detailed exposition of the impact origin of virtually all lunar craters and basins, see Wilhelms, 1987). The most detailed ranking of the relative ages of multi-ring basins on the Moon is the one by Wilhelms (1987) and is adopted here (Tables 2.1, 2.2).

The existence of most basins is established by detailed photogeologic mapping; supplementary evidence in the form of topographic information (to demonstrate the presence of regional lows) and geophysical data (e.g., mascons or zones of crustal thinning) provide additional support for the presence of a basin. Basins are recognized by the following morphological criteria, ordered in terms of decreasing reliability: (1) the presence of strongly expressed concentric scarps, of which the Cordillera ring of the Orientale basin is the best example (Figure 2.7g); (2) arcuate chains of massifs, either isolated or associated with textured ejecta deposits; (3) arcuate chains of massifs in

Table 2.1 *Lunar multi-ring basins*

	Basin[a]	Center	Relative Age[b]	Ring diameters (km)[c]
1	Orientale	19 S, 95 W	I	320, 480, 620, **930**, 1300, 1900
2	Imbrium	35 N, 17 W	I	550, 790, **1160**, 1700, 2250, 3200
3	Hertzsprung	2 N, 128 W	N	150, 255, 380, **570**
4	Serenitatis	26 N, 18 E	N	410, 620, **920**, 1300, 1800
5	Crisium	18 N, 59 E	N	360, 540, **740**, 1080, 1600
6	Humorum	24 S, 39 W	N	210, 340, **425**, 570, 800, 1195
7	Humboldtianum	59 N, 82 E	N	250, 340, 460, **650**, 1050, 1350
8	Korolev	4 S, 158 W	N	220, **440**, 590, 810
9	Moscoviense	26 N, 148 E	N	140, 220, 300, **420**, 630
10	Mendel–Rydberg	50 S, 94 W	N	200, 300, **420**, 630
11	Nectaris	16 S, 34 E	N	240, 400, 620, **860**, 1320
12	Apollo	36 S, 151 W	pN	240, **480**, 720
13	Grimaldi	6 S, 68 W	pN	230, 300, **440**
14	Smythii	2 S, 87 E	pN	260, 370, 540, **740**, 1130
15	Coulomb–Sarton	52 N, 123 W	pN	160, 250, **440**, 670
16	Keeler–Heaviside	10 S, 162 E	pN	325, **500**, 750, 1000
17	Ingenii	43 S, 165 E	pN	165, **315**, 450, 660
18	Balmer	15 S, 70 E	pN	260, **500**, 750, 1000

Notes: [a]Listed in reversed order of relative age (e.g., Orientale is the youngest, Balmer is the oldest of this group) according to Wilhelms (1987). See Figure 2.9 for basin location.
[b]Stratigraphic relative ages: I – Imbrian, N – Nectarian, pN – pre-Nectarian
[c]Diameter in boldface is the most topographically prominent ring, here considered to be the main topographic rim of the basin
Source: Ring diameters from Pike and Spudis (1987).

association with wrinkle ridges within the maria; (4) circular arrangements of complex, wrinkle-ridge systems within the maria; and (5) isolated massifs, scarps, and ridges that collectively occur at more or less constant radial distances from a central point.

The diversity of ring expression observed in lunar basins is a function of both relative age and ring position. For the most ancient basins, subsequent deposition of geological units and erosion by impact degradation obscure ring prominence and morphology. Moreover, the viscous relaxation of basin topography can also reduce the expression of surface relief (e.g., Baldwin, 1987a). However, even the freshest basins may display obscure, poorly developed ring systems (e.g., outer rings of Orientale; Hartmann and Kuiper, 1962; Pike and Spudis, 1987). Thus, workers frequently find that their tabulations of ring diameters and positions differ, in some cases to the point of disagreement as to whether one or two basins are present (e.g., the Imbrium–Procellarum basin controversy; see Chapter 7).

Lunar basins that show fewer than three rings (Table 2.2) are the most ancient and heavily degraded features. On the basis of their relatively large sizes, many of these basins were probably multi-ring features when they originally formed. It is

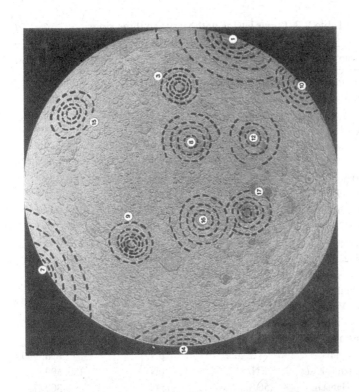

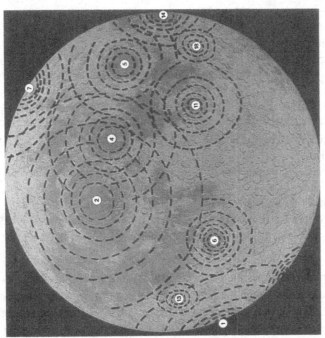

Figure 2.9 The location and rings of multi-ring basins on the Moon displaying three or more rings. See Table 2.1 for basin centers and ring diameters. Base is Lambert equal-area projection by the U.S. Geological Survey.

Table 2.2 Lunar basins with less than three rings

	Basin[a]	Center	Relative age[b]	Ring diameters (km)[c]
19	Schrödinger	76 S, 134 E	I	150, **320** *
20	Bailly	67 S, 68 W	N	150, **300** #
21	Sikorsky–Rittenhouse	68 S, 111 E	N	**310** #
22	Mendeleev	6 N, 141 E	N	140, **365** *
23	Milne	31 S, 113 E	pN	125, 262 #
24	Freundlich–Sharonov	18 N, 175 E	pN	**600** #
25	Birkhoff	59 N, 147 W	pN	150, **325** *
26	Planck	58 S, 136 E	pN	160, **325** *
27	Schiller–Zucchius	56 S, 45 W	pN	175, **335** *
28	Amundsen–Ganswindt	81 S, 120 E	pN	175, **335** *
29	Lorentz	34 N, 97 W	pN	170, **365** *
30	Poincaré	57 S, 164 E	pN	160, **325** *
31	Lomonosov–Fleming	19 N, 105 E	pN	**620** #
32	Nubium	21 S, 15 W	pN	**690** #
33	Mutus–Vlacq	51 S, 21 E	pN	**690** #
34	Tranquillitatis	7 N, 30 E	pN	**700**, 950
35	Australe	52 S, 95 E	pN	550, **880** #
36	Fecunditatis	4 S, 52 E	pN	**690**, 990 #
37	Al-Khwarizmi–King	1 N, 112 E	pN	250, **590**
38	Pingre–Hausen	56 S, 82 W	pN	300 (?)
39	Werner–Airy	24 S, 12 E	pN	**500** #
40	Flamsteed–Billy	7 S, 45 W	pN	320, **570**
41	Marginis	20 N, 84 E	pN	**580** #
42	Insularum	9 N, 18 W	pN	**600**, 1000
43	Grissom–White	44 S, 161 W	pN	600 (?)
44	Tsiolkovsky–Stark	15 S, 128 E	pN	**700** #
45	South Pole–Aitken	56 S, 180 E	pN	1900, **2600**

Notes: [a]Listed in order of decreasing relative age according to Wilhelms (1987). See Figure 2.10 for basin location.

[b]Relative stratigraphic age: I – Imbrian, N – Nectarian, pN – pre-Nectarian, according to Wilhelms (1987).

[c]Underlined diameter is main topographic rim as here interpreted (where discernable.

Sources: Ring diameters marked with an asterisk (*) are from Pike and Spudis (1987), diameters marked with a sharp sign (#) are from Wilhelms (1987), all others are my own measurements.

possible that future data will permit the recognition of additional ring structures that presently cannot be seen on even the best photographs. Thus, the data presented in Tables 2.1 and 2.2 should be considered tentative; additional exploration of the Moon will probably lead to the recognition of more multi-ring basins. The next five chapters of this book consider the geology of five multi-ring basins in some detail; we will then return to the problems of basin formation and development.

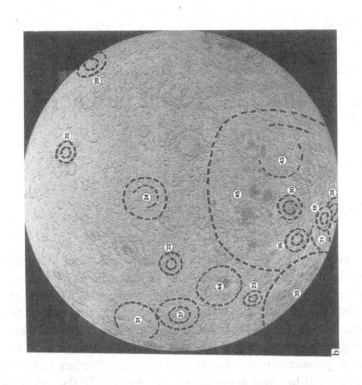

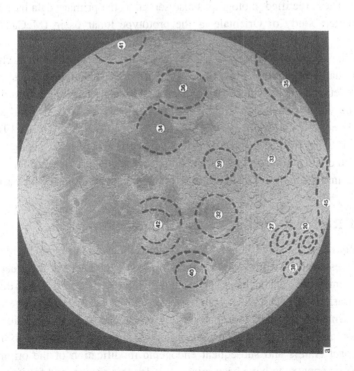

Figure 2.10 The location and rings of basins on the Moon having less than three rings. See Table 2.2 for basin centers and ring diameters. Lambert equal-area projection by the U.S. Geological Survey.

3

The "archetype" basin: Orientale

The Orientale basin (Figure 3.1) is located on the western limb of the Moon; most of the basin and its deposits extend over the lunar far side. The name "Orientale" (eastern) is derived from the old astronomical practice of displaying telescopic photographs of the Moon with south at the top and east-west convention derived from terrestrial coordinates. The first studies of the Orientale basin utilized Earth-based telescopic photographs; because of the basin's location on the extreme limb, these photographs were geometrically rectified at the Lunar and Planetary Laboratory (University of Arizona) to create a vertical viewing perspective (Hartmann and Kuiper, 1962). These rectified photographs also served as the primary data base for an early geological study of Orientale as the prototype lunar basin (McCauley, 1968).

Because of the spectacular success of the Lunar Orbiter spacecraft, particularly a series of high-quality photographs by Orbiter IV that provide contiguous coverage, the number of detailed geological descriptions of the Orientale basin increased dramatically in the years immediately following the Apollo missions (Hartmann and Wood, 1971; Head, 1974a; Moore *et al.*, 1974; McCauley, 1977; Scott *et al.*, 1977; Spudis *et al.*, 1984b). In this chapter, I review the regional and local geology of Orientale and, in conjunction with data from photogeology and remote sensing, integrate these data into a geological model for the formation and evolution of the basin.

3.1 Regional geology of the Orientale impact site

The Orientale basin is sparsely filled by mare basalt and located in rugged highland terrain on the western limb of the Moon (Figure 3.1). The thickness of the crust in this region is somewhat greater than usual, ranging from 90 to over 120 km thick (Bills and Ferrari, 1976). Orientale is the youngest large basin (Howard *et al.*, 1974) and it formed when the Moon had already developed a thick, rigid lithosphere (Solomon and Head, 1980). Thus, its geological setting is analogous to that of smaller impact craters and subsequent endogenic modification of the original basin morphology appears to have been minor. It is for this reason, and for its lack of burial by younger geological units, that Orientale has been used as a model for the process of basin formation on the Moon.

The impact site for Orientale resembled the rugged, unflooded terra now well exposed to the southwest of the basin. Several older basins have been identified in the Orientale region (Hartmann and Wood, 1971; Schultz and Spudis, 1978; Spudis

Figure 3.1 Basin-centered view of the Orientale basin. The innermost ring (1) is about 320 km diameter and is defined by a discontinuous shelf bounding the inner basin mare fill. The inner Rook ring (2) is 480 km diameter and is made up of rugged massifs. The outer Rook ring (3) is 620 km diameter, composed of massifs, and shows a quasi-polygonal outline. The Cordillera ring (4) is 930 km diameter and marks the topographic rim of the basin. Two additional rings outside the Cordillera ring are illustrated in Figure 3.9. Basin secondary chains may occur near the basin rim (e.g., Vallis Bouvard- B). LO IV-187 M.

et al., 1984b). Although most of these structures lie far outside its topographic rim, Schultz and Spudis (1978) identified a system of pre-basin pyroclastic vents (Figure 3.2) within the rim of Orientale that, along with preserved elements of an outer ring, may represent a pre-Orientale two-ring basin that influenced the shape of the outer and intermediate rings of Orientale. Moreover, many smaller craters that have been mapped within the main ring (Cordillera Mountains) appear to be older than Orientale (Head, 1974a; King and Scott, 1978; Schultz and Spudis, 1978; Scott *et al.*, 1977). These features provide important clues to the probable location of the basin excavation cavity, as I describe below.

Figure 3.2 Zond 8 photograph of the Orientale basin under high solar illumination. Low-albedo ring (arrows) mark the location of a preserved series of pre-basin volcanic vents extending inside the outer Rook ring (Schultz and Spudis, 1978).

Table 3.1 *Geologic units of the Orientale basin*

Basin unit	Characteristics	Interpretation	Head, 1974a	Moore *et al.*, 1974	Scott *et al.*, 1977
Smooth plains	High-albedo plains; mantle and embay topography	Clast-poor impact melt	Plains facies	Central basin material	Plains materials
Fractured plains	High-albedo rough surface, fractured; mantles topography	Clast-rich impact melt	Corrugated facies	Central basin material	Maunder Formation
Knobby deposits	Hummocky knobs set in rough matrix; occurs mainly between Cordillera and outer Rooks, may locally extend beyond Cordillera	Basin ejecta	Domical facies	Knobby basin material	Montes Rook Formation
Textured deposits	Concentric and radially lineated texture; occurs outside Cordillera ring	Basin ejecta blanket	Textured ejecta	Concentric and radial facies	Hevelius Formation

3.2 Orientale morphology and geological units

Orientale geological units (Table 3.1; Figure 3.3) are delineated by spatial relation to the basin rings and by morphological texture. Most workers agree on the definition of geological units (dominantly distinguished by morphological properties) but differ in their interpretation of unit origins and the significance of unit distribution.

3.2.1 Basin interior units

Excluding the mare basalt deposits, which post-date the basin, the interior units of the Orientale basin are subdivided on the basis of morphology into three units (Table 3.1). The innermost unit is composed of smooth, moderately high-albedo plains that embay and overlie the inner rings (Figure 3.4); the plains have been named the smooth facies of the Maunder Formation by Scott *et al.* (1977). These smooth plains grade laterally outward from the basin center into a rough-textured mantling deposit, alternatively termed the Maunder Formation, fissured facies (Moore *et al.*, 1974) or corrugated facies (Head, 1974a). These mantling deposits appear to overlie the

Figure 3.3 Geological sketch map of the Orientale basin; geology after Scott *et al.* (1977), McCauley (1977), McCauley *et al.* (1981), and Wilhelms *et al.* (1979). Base is stereographic projection centered on 19°S, 95°W (Orientale basin center).

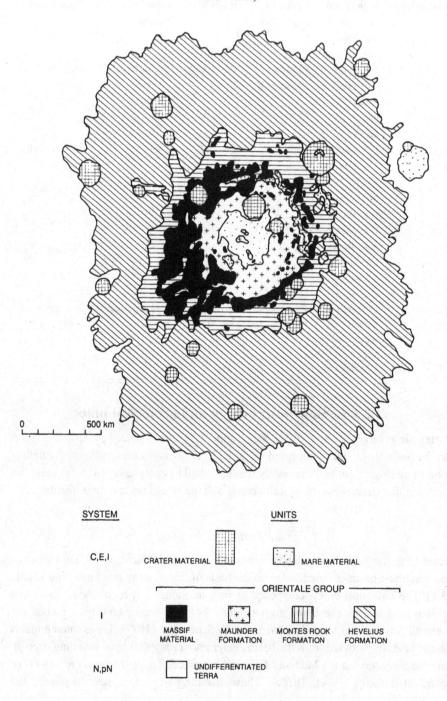

SYSTEM	UNITS

C,E,I   CRATER MATERIAL   MARE MATERIAL

─────────────────────── ORIENTALE GROUP ───────────────────────

I   MASSIF   MAUNDER   MONTES ROOK   HEVELIUS
    MATERIAL   FORMATION   FORMATION   FORMATION

N,pN   UNDIFFERENTIATED
       TERRA

0   500 km

rugged basin floor and are similar in morphology and distribution to rough-textured deposits seen on the floors of fresh lunar craters, which have been interpreted to be impact melt deposits (Shoemaker *et al.*, 1968; Howard and Wilshire, 1975). In a like manner, the Maunder Formation is interpreted as Orientale impact melt by most investigators (Head, 1974a; Moore *et al.*, 1974; McCauley, 1977; Scott *et al.*, 1977). These deposits appear to have remained at least semi-molten for a considerable length of time, because lava-like channels appear to have carried melt into the center of the basin (Greeley, 1976). Long cooling times are also suggested by the morphol-

Figure 3.4 Inner basin geology of Orientale. Smooth plains (S) and fissured deposits (F) make up the Maunder Formation. These deposits grade laterally outward into the knobby Montes Rook Formation (R). North at top, framelet width 12 km. LO IV-195 H1.

ogy of unusual large craters such as Kopff, which may owe their distinctive appearance to impact into a still-cooling melt mass (Wilhelms and McCauley, 1971; Guest and Greeley, 1977). The impact melt sheet of the Orientale basin provides a morphological equivalent to the melt sheets of other lunar basins that are now buried by mare basalt deposits. The Maunder Formation appears to terminate at the Outer Rook ring (3 in Figure 3.1) of Orientale.

The next unit, which is primarily between the Outer Rook and Cordillera scarp, is

Figure 3.5 The Montes Rook Formation. This knobby, inner basin deposit may locally extend beyond the Cordillera scarp (C) and overlie the outer basin Hevelius Formation (H). Small pools of material, probably impact melt, occur within the knobby units (P). North at top, framelet width 12 km. LO IV-187 H3.

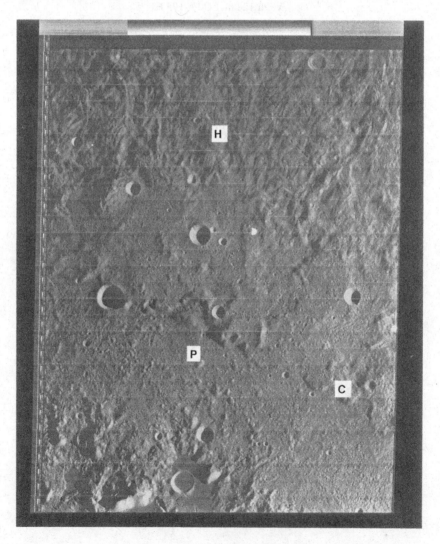

the hummocky Montes Rook Formation (termed knobby materials by Moore *et al.*, 1974 and domical facies by Head, 1974a; see Table 3.1). It consists of widely spaced knobs, on the order of 10 km in size, set in an undulating to hummocky matrix (Figure 3.5). Although the Montes Rook Formation is concentrated between the outer and intermediate rings of Orientale, in several places it appears to extend beyond the outer (Cordillera) scarp and may overlie the textured ejecta deposits found beyond the basin rim (McCauley, 1977). This relation has been cited as evidence that the Montes Rook Formation must be a petrologically distinct facies of ejecta, emplaced late in the cratering process (McCauley, 1977; Scott *et al.*, 1977). Alternatively, it has been suggested that the Montes Rook Formation is a structural

Figure 3.6 The Hevelius Formation. Textured deposits of the Orientale basin overlie pre-Orientale craters in this region (e.g., Riccioli – R). Dune-like morphology is seen in the basin-facing walls of such craters (D). North at top, framelet width 12 km. LO IV-173 H3.

texture, imposed upon basin ejecta during the seismic disturbances associated with the megaterrace formation of the Cordillera scarp (Head, 1974a). I believe this latter interpretation to be unlikely, as not only does the Montes Rook material extend outside the Cordillera, the Hevelius Formation (discussed below) occurs *inside* the basin rim (Figure 3.3); these relations suggest that the two units are different facies of ejecta, and not surface textures imposed by seismic shaking, which would have affected all units in the basin.

There is little unequivocal evidence to suggest that impact melt makes up a large proportion of the Montes Rook Formation (McCauley, 1977), although the inclusion of melt masses within primarily clastic ejecta is possible. Some regions of the Montes Rook Formation exhibit small pools of fissured material (Figure 3.5) that are probably ponds of impact melt that have segregated from the main melt sheet by low-energy ejection from the basin cavity. The cause of the knobby texture of this unit is unknown, but units of similar morphology are found around other basins, as will be discussed in Chapters 4 and 7.

3.2.2 Basin exterior units

Textured ejecta outside of the Cordillera scarp of the Orientale basin were recognized during telescopic mapping of the Moon (McCauley, 1967, 1968) and a type area was defined near the crater Hevelius, hence the unit has been formally named the Hevelius Formation (Wilhelms, 1970, 1987; Table 3.1). It consists of hummocky, radially lineated, and swirl-textured deposits (Figure 3.6) that may extend to distances of about one basin radius (about 500 km) beyond the Cordillera scarp. The Hevelius Formation is varied in thickness and there are traces of many pre-Orientale buried craters, even near the main basin rim. A pronounced concentric texture near the Cordillera ring (Moore *et al.*, 1974) suggests that post-depositional slumping was important in the development of ejecta deposits near the basin topographic rim.

In many places, the surface texture of the Hevelius Formation suggests that some type of ground flow occurred during the final stages of ejecta deposition (Moore *et al.*, 1974; Chao *et al.*, 1975). Deceleration lobes occur on the downrange walls of pre-basin craters, imparting a wormy texture to the continuous ejecta (Figs. 3.6, 3.7). Low-albedo fissured deposits that resemble the fissured facies of the Maunder Formation also are found in isolated areas within this sequence (Moore *et al.*, 1974; Schultz and Mendenhall, 1979). This material may be ejected melt masses entrained within the clastic ejecta that make up the bulk of the Hevelius Formation. Moreover, flow lobes of high-albedo material appear to have segregated from the textured ejecta (Figure 3.7) and were emplaced in part by leveed channels (Moore *et al.*, 1974; Schultz, 1976a; Wilhelms *et al.*, 1980; Wilhelms, 1987). Oberbeck (1975) argued that these lobes may be flows of clastic debris, but I believe that their morphology is more consistent with an impact melt origin.

The textured ejecta deposits of the Orientale basin grade outward into smooth to undulating deposits of highland plains. Light plains on the Moon have been the source of considerable controversy since the Apollo 16 mission demonstrated that they are composed of impact breccias. Chao *et al.* (1975) maintain that the light plains surrounding Orientale consist of primary basin ejecta; moreover, they contend that this smooth facies of Orientale ejecta may have been deposited as a debris blanket over the entire Moon. Moore *et al.* (1974) further suggest that Orientale ejecta may have been concentrated at the antipodal region of the Moon, near Mare Marginis, which resulted in an unusual furrowed texture seen in this region (Wilhelms and El-Baz, 1977). Alternatively, this furrowed texture may result from concentration of seismic waves antipodal to the basin produced during the Orientale impact (Schultz and Gault, 1975).

Figure 3.7 The distal margins of the continuous deposits of the Orientale basin. Basin secondary impact craters are abundant here (S). Flow lobes (F) are contiguous with continuous deposits and may consist of ejected impact melt. North at top, framelet width 12 km. LO IV-180 H1.

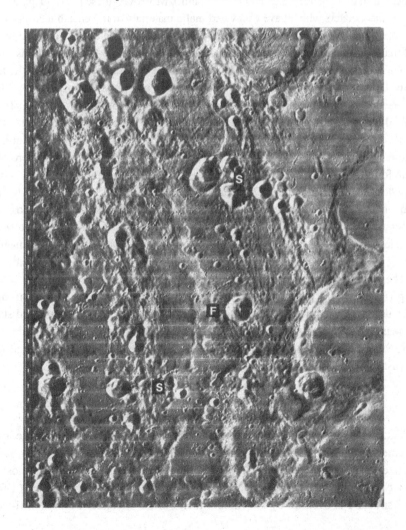

Oberbeck (1975) argues that the light plains near the distal ends of basin ejecta blankets are nearly all ejecta from secondary craters and thus of local provenance; he estimates that primary components make up less than 10–20% of the deposits. Schultz and Gault (1985) argue that these estimates of primary ejecta are dependent upon the assumption of single-body impactors for the observed secondary craters and that clouds of dispersed ejecta could result in much higher values of basin primary ejecta. Wilhelms (1987) asserts that the Oberbeck estimates are model-dependent and that small but reasonable changes in the assumed initial conditions can drastically affect the mixing ratios. I consider these objections well taken and contend that for at least the light plains near the distal margins of the textured ejecta, Orientale primary ejecta may constitute at least 50% and perhaps as much as 75% of the plains materials (Schultz and Gault, 1985).

The light plains surrounding the periphery of the textured ejecta have filled many craters to shallow depths (e.g., crater Wargentin). Moreover, these plains display dark-halo impact craters which have excavated mafic materials from beneath them (Schultz and Spudis, 1979; Hawke and Bell, 1981). The plains ejecta in this region have buried old mare basalt deposits which partially filled many of the pre-Orientale basins found in this region. The dark-halo craters imply that mare volcanism was active at the time of the Orientale impact and was probably an ongoing process at the basin impact site.

Secondary craters from the Orientale impact (Figure 3.7) are widely distributed around the distal margins of the Hevelius Formation (Wilhelms, 1976a; Scott *et al.*, 1977). Moreover, chains of secondary craters occur close to the basin rim within the textured ejecta deposits, such as Vallis Bouvard (Figure 3.1). These craters probably were formed by the ejection of material at low angles, accompanying the ground-hugging debris surge (Oberbeck, 1975) that emplaces basin ejecta. Some secondary craters in the outer regions of the Orientale deposits display low-albedo, fissured deposits on the crater floor; these deposits may represent melt ejecta from the basin (Schultz and Mendenhall, 1979) that could have resulted from the ejection of mare basalt magmas during basin excavation (Schultz and Mendenhall, 1979; Ryder and Spudis, 1980).

The relations among Orientale units are summarized in the geological sketch map (Figure 3.3). These relations, at the best preserved basin on the Moon, have been used to guide our understanding of other lunar basins in more advanced states of degradation. However, Orientale displays several peculiarities (discussed in the last section) that indicate that its use as the basin "archetype" may be inappropriate, at least for some aspects of the basin problem.

3.3 Rings and basin structures

Orientale displays at least three and possibly as many as six concentric rings (Figures 3.1, 3.8). The innermost, shelf-like ring (320 km diameter) is partly buried by later mare deposits (Figure 3.4). This ring appears to encompass the positive gravity anomaly (mascon) seen in the orbital gravity data (Sjogren *et al.*, 1974). This structure is overlain by the smooth facies of the Maunder Formation (impact melt sheet). The next ring consists of rugged massif segments and has been named the inner Montes Rook ring (480 km diameter). It is at this position within the basin that the transition between smooth and fissured members of the Maunder Formation occurs, possibly a result of the thinning of the melt sheet moving outward from the center of the basin. This ring has been interpreted both as the rim of the transient cavity (Floran and Dence, 1976) and as a central uplift analogous to a crater central peak (Head, 1974a; Moore *et al.*, 1974; McCauley, 1977). In the nested crater model for basin ring formation, this ring represents the surface expression of the crust–mantle boundary of the Moon (Hodges and Wilhelms, 1978).

The next ring is the outer Montes Rook, about 620 km in diameter. At this posi-

Figure 3.8 Ring interpretation for the Orientale basin. See Table 2.1 for ring diameters. Base is a stereographic, basin-centered projection. From Pike and Spudis (1987).

tion, there is an abrupt transition between the Maunder Formation and the Montes
Rook Formation (Figure 3.4). The outer Rook ring is composed of rugged massif
elements (Figures 3.1, 3.3), but displays considerable departure from the normal
circular plan seen in the inner Rook ring, particularly at the southwestern edge of

Figure 3.9 The two outer rings of the Orientale basin. "IV" denotes the 930 km
diameter Cordillera ring, "V" is the 1300 km diameter Rocca ring (Hartmann and
Kuiper, 1962). "VI" is a large (1900 km diameter), discontinuous ring described
by Pike and Spudis (1987). From Pike and Spudis (1987); LO IV-167 M.

the ring. This area is the location of a pre-Orientale, two-ring basin described by Schultz and Spudis (1978). If this older basin exists, as I believe it does, the transient cavity for the Orientale basin must lie within the outer Rook ring to preserve this vestige of pre-basin structure.

The topographic rim of the Orientale basin is delineated by the Cordillera Mountains (930 km diameter). This ring exhibits simple, scarp-like morphology as opposed to the blocky massifs displayed by the two Rook rings. At this position, the deposits abruptly change in morphology from Montes Rook Formation within the basin topographic rim to the textured ejecta of the Hevelius Formation outside it. In some places, the Montes Rook Formation extends beyond the Cordillera for

Figure 3.10 Recognizable pre-basin structures in the Orientale region. Dot–dash lines are craters mapped by Head (1974a), dashed lines are craters recognized by King and Scott (1978), and solid lines are features mapped by Schultz and Spudis (1978). Thin line denotes the Cordillera scarp (rim of Orientale basin). After Spudis *et al.* (1984b).

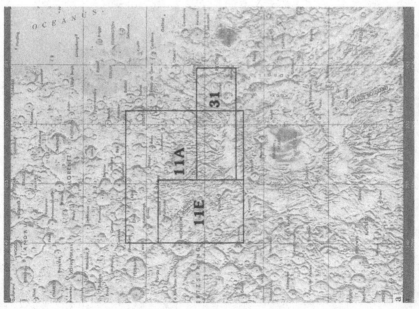

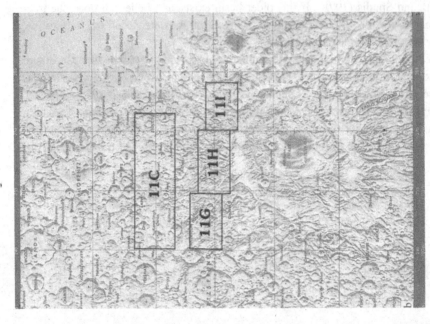

Figure 3.11 Index maps for regions of Orientale basin continuous deposits for which mixing models were performed (Table 3.2). Boundaries of regions are defined in Metzger et al. (1977).

distances of less than a few tens of kilometers. The Cordillera scarp is very circular in plan on the well photographed eastern side of the basin, but becomes extremely irregular on the southwestern and western side (Figure 3.8).

In their initial study of Orientale, Hartmann and Kuiper (1962) identified a large ring exterior to the Cordillera scarp. Subsequent studies (e.g., Moore *et al.*, 1974; Head, 1974a) have tended to ignore this outer ring. However, Pike and Spudis (1987) reexamined the rings of all lunar basins and found evidence not only for this large, exterior ring of Hartmann and Kuiper (1962), but also for an additional ring outside of this one (Figures 3.8, 3.9). These two exterior rings have diameters of 1300 and 1900 km, respectively, are mantled by Orientale ejecta (Hevelius Formation), and appear to have subdued, scarp-like morphology (Figure 3.9). Even though Orientale is one of the freshest, best preserved lunar basins, these outer rings are very subtle and careful mapping was required to establish their presence. This subdued expression suggests that large rings exterior to the basin rim may exist around many basins, but cannot be identified.

Hartmann and Kuiper (1962) noted that the ring diameters of Orientale increase outward in relative size by a constant value; they suggested a spacing factor of about 1.4, i.e., each ring is about 1.4 times the diameter of the ring just inside of it. Later, Fielder (1963) argued that this factor is very close to the square root of two ($\sqrt{2} D$). Hartmann and Wood (1971) systematized this relation to include all lunar basins. Most recently, Pike and Spudis (1987) examined 67 multi-ring basins on three planets and concluded that the "square root of two" spacing rule holds for the Moon, Mercury, and Mars, and possibly, for Earth and some icy satellites of the Jovian planets (Pike, 1985; Moore *et al.*, 1984). Orientale displays the $\sqrt{2} D$ spacing for all six of its rings (Pike and Spudis, 1987) and, in fact, fits the "ideal" spacing model better than does any basin in the Solar System. This ring spacing has significant implications for models of ring genesis and will be discussed in Chapter 8.

The basin rings of Orientale depart from a circular plan by interaction with many older structures. The pre-Orientale, two-ring basin and its pyroclastic vent system (Figure 3.2) have interfered with and probably caused the disruption of both the outer Rook and Cordillera rings. A sketch map synthesizing the decipherable pre-basin features in this region is shown in Figure 3.10. Of particular interest is a large crater, on the western rim of the basin near the crater Lowell, that appears to be cut in half by the Cordillera scarp (Schultz and Spudis, 1978). If this interpretation is correct, it suggests that the Cordillera ring of the basin is a fault scarp. The presence of both a fault-disrupted crater and an interior two-ring basin further implies that the basin transient cavity cannot be represented by the Cordillera scarp and possibly, not even the outer Rook ring. If either of these basin rings were the true rim of the Orientale crater, the pre-basin features could not have been preserved.

3.4 Remote sensing of Orientale basin deposits

Geochemical data are available for a large portion of the Orientale ejecta blanket. Both Apollo 15 and 16 ground tracks pass over the northern rim of the basin and

Table 3.2 *Chemical composition and mixing model results for
Orientale basin deposits*

Region	Geologic unit	Fe (wt. %)	Ti (wt. %)	Al (wt. %)	Th (ppm)	FAN %	GABAN %	ANGAB %	LFKM %	17MB %
11 A	Hevelius Fm.	5.0	0.7	13.2	0.37	25	5	54	5	11
11 C	Distal Hevelius Fm.	4.7	0.5	13.4	0.48	—	49	44	1	6
11 E	Hevelius Fm.	5.0	0.9	13.2	0.27	40	1	43	2	14
11 G	Hevelius Fm.	2.3	0.4	15.4	0.29	63	—	32	2	3
11 H	Hevelius Fm.	4.2	0.2	13.8	0.33	51	—	40	4	5
11 I	Hevelius Fm.	6.1	0.6	12.2	0.51	—	26	60	4	10
31	Hevelius/ Montes Rook Fms.	4.5	0.6	13.6	0.41	45	—	40	6	9

Source: Regions defined in Metzger *et al.* (1977). Chemistry from Davis (1980) and Metzger *et al.* (1977); Al values estimated from regression of Fe concentrations (Spudis *et al.*, 1988a). Mixing model results from Spudis *et al.* (1984b).

orbital coverage extends from the border of Oceanus Procellarum in the east to the discontinuous Orientale deposits west of Hertzsprung (Figure 3.11). Because these basin deposits are virtually unmodified by subsequent mare flooding, the composition of the Orientale ejecta blanket should be the most nearly original and thus best characterized of any basin on the Moon. Unfortunately, only gamma-ray data exist for this area because it was in darkness during the Apollo missions. Therefore, only the concentrations of the elements Fe, Ti, and Th are available for the ejecta blanket; however, it is possible to estimate regional Al concentrations from the Fe data based on a nearly perfect, Fe–Al inverse correlation for lunar soils (Schonfeld, 1977; Spudis *et al.*, 1988a). Thus, mixing model studies can be performed using these elements, but the sparse data base generally provides poorer fits than mixing models for other basin ejecta, where X-ray data are available.

Regional composition of the Orientale ejecta deposits is consistent with a composition rich in anorthosite (Table 3.2). This area also contains some of the lowest concentrations of Th seen on the lunar surface, in general, less than 0.5 ppm (Metzger *et al.*, 1977). The composition of the circum-Orientale highlands suggests that the levels of KREEP in Orientale basin ejecta is the lowest seen for any lunar basin, which constrains the depth of basin excavation to upper crustal levels (Spudis, 1982), as will be discussed below.

The results of mixing models performed on these geochemical data are presented in Table 3.2. The observed data can be modeled best as a three-component mixture of anorthosite (FAN), gabbroic anorthosite (GABAN), and anorthositic gabbro (ANGAB). The minor component of mare basalt is required primarily to account for excess Ti, as few lunar highland samples contain any appreciable amounts of this element. The low abundance of low-K Fra Mauro basalt (LKFM) is particularly interesting because this composition is associated almost exclusively with basin impact melts collected at the Apollo 15 and 17 landing sites and appears to be derived from deep crustal levels (Ryder and Wood, 1977; Spudis and Davis, 1986). The paucity of the LKFM composition in Orientale deposits again suggests very little material derived from lower crustal levels is present.

Mixing model results imply that anorthosite is much more abundant than norite in Orientale deposits and that virtually no KREEP is present. This strengthens the interpretation that little or no materials derived from lower levels of the crust are present in the basin ejecta. The high amounts of calculated mare basalt in regions

Figure 3.12 Index photograph showing locations of near-infrared spectra of Orientale basin deposits. Numbers refer to spectra presented in Figure 3.13. After Spudis *et al.* (1984b).

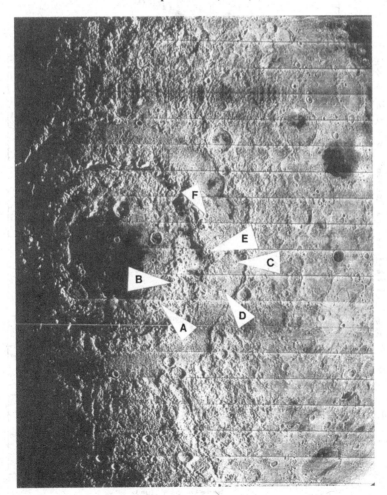

11A, 11E, and 11I probably result from excess Ti and Fe in these regions (Table 3.2). A highland rock type containing relatively high levels of Ti would change these model results and it appears that at least some pristine norites do contain relatively high amounts of Ti (Ryder and Norman, 1978a). The use of these norites as end members still gives abundances of less than 10 percent and does not invalidate the conclusion that mafic components are virtually absent in Orientale ejecta.

Compositional information from spectral studies for the Orientale basin largely

For legend see next page.

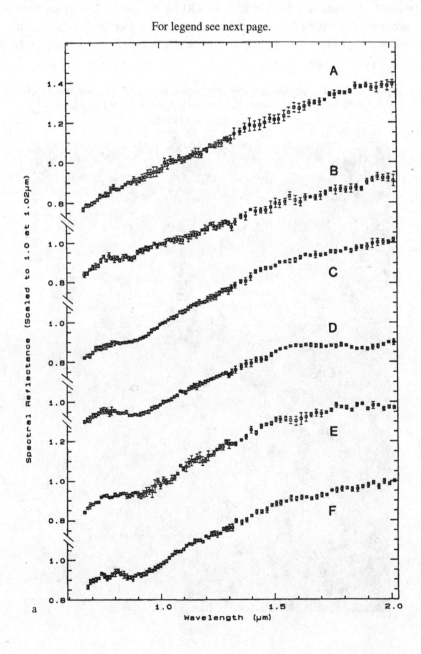

a

substantiates the inferences drawn from the orbital chemical data. Orientale's position near the lunar limb makes it a difficult target for Earth-based observations. However, Spudis *et al.* (1984b) obtained 11 near-infrared spectra of targets within and near the Orientale basin; six selected spectra were obtained for parts of the

Figure 3.13 Spectra for six features in the Orientale basin; see Figure 3.12 for locations. (a) Raw spectra. (b) Continuum removed spectra. See text for discussion. From Spudis *et al.* (1984b).

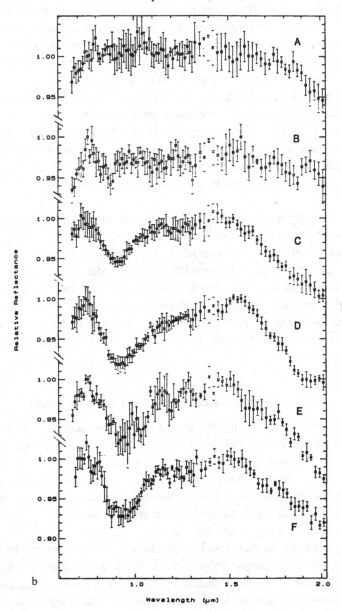

inner rings of the basin and small craters in the Montes Rook Formation (Figures 3.12, 3.13). In general, Orientale deposits appear similar in composition to those of the Apollo 16 highlands, ranging from anorthositic norite to noritic anorthosite (C–F, Figure 3.13), in agreement with the orbital chemical data.

Two spectra for the inner Rook ring (A and B, Figures 3.12, 3.13) show no evidence for the 1 micron absorption band characteristic of orthopyroxene, a common, though minor, component of most lunar highlands soils. Spudis *et al.* (1984b) interpreted these two spectra as indicating the presence of pure anorthosite within the basin rings; this was the first discovery of such outcrops on the Moon. Subsequent studies have shown that anorthosite crops out in several additional places on the Moon, notably within the Nectaris and Humorum basins (Spudis *et al.*, 1989; Spudis *et al.*, 1992), but nearly always in association with the *inner* rings of lunar basins (Hawke *et al.*, 1991), a relation as yet unexplained.

Data from Earth-based spectral studies and the Apollo orbital geochemical experiments indicate that Orientale basin deposits are dominantly of anorthositic composition. No evidence is found for mafic (e.g., norite) or ultramafic (e.g., dunite) rock types within the basin or its deposits. The geochemical data suggest that low-K Fra Mauro basalt cannot constitute more than about 5 percent of the ejecta. No systematic radial or concentric variations in the composition of Orientale basin ejecta are observed.

These remote-sensing results have important implications for the depth of excavation of the Orientale basin. The impact occurred into a thick, highland crust where the lithosphere was also relatively thick (sec. 3.1). From photogeologic evidence, the basin ejecta have been little modified after emplacement. Anorthosite and anorthositic rocks make up the bulk of the ejecta and the virtual absence of a mafic and KREEP-rich component suggests that lower levels of the crust were not sampled. Therefore, the basin impact can have excavated only about the upper half of the crust, in this region, to depths shallower than about 50–60 km.

3.5 Orientale ejecta at Apollo landing sites?

Chao *et al.* (1975) proposed that the impact forming the Orientale basin deposited ejecta over the entire surface of the Moon. If this hypothesis is correct, it might be expected that Orientale ejecta would have been sampled at the Apollo and Luna highland landing sites (Figure 1.1). The results of the remote-sensing studies suggest a uniform, anorthositic composition for this material at large radial distances from the basin and, if Orientale ejecta are to be found at any site, they may be expected to have a very similar composition.

The highland landing site closest to the Orientale basin is the Apollo 14 site at Fra Mauro, chosen to sample Imbrium basin ejecta. Study of the photographs of this region reveal no Orientale secondary craters or deposits attributable to that basin. Moreover, the samples returned from the Fra Mauro region are extremely rich in KREEP, basaltic in bulk composition, and anorthosite is sparse to absent (Swann *et al.*, 1977; Simonds *et al.*, 1977); these characteristics are exactly the reverse of what is seen in the ejecta from the Orientale basin. I conclude that no Orientale ejecta were sampled at the Apollo 14 landing site.

Most attention in the search for Orientale material at the Apollo sites has been concentrated on Apollo 16, mainly because of the supposed contemporaneity between the Cayley plains at this site in the Descartes region and the smooth plains deposits associated with Orientale ejecta deposits (Eggleton and Schaber, 1972; Chao *et al.*, 1975; Hodges and Muehlberger, 1981). The Apollo 16 landing site is indeed anorthosite-rich and very similar to that of the Orientale ejecta seen from orbit. However, regional variations in chemistry around the Descartes site suggest that this region of the Moon is a distinct geochemical province and evidence for a regional blanket of uniform ejecta is absent (Spudis and Hawke, 1981). There is no photogeologic evidence for Orientale secondary craters or deposits in the Descartes region (Head and Hawke, 1981).

Any search for Orientale ejecta at the Apollo 16 site must, therefore, concentrate on specific samples that may be exotic to the site and that have crystallization ages contemporaneous with the Orientale impact. One candidate sample is the impact melt rock 68415, which has a composition of anorthositic gabbro and a crystallization age of about 3.82 Ga (Deutsch and Stöffler, 1987). This rock was collected at the site from a boulder ejected from shallow levels of the Cayley plains (Reed, 1981), where Orientale ejecta should be concentrated (Chao *et al.*, 1975). However, the composition of 68415 is very similar to the background composition of the local materials; chemically, it can be modeled as a mixture of Descartes and Cayley soils. Therefore, while an Orientale origin for this sample cannot be excluded, the petrology of the rock suggests it may be of local origin, possibly from one of the many post-Imbrium local craters.

There is no petrologic or geochemical evidence to support the claim of Chao *et al.* (1975) that Orientale ejecta were deposited as a regional debris blanket everywhere on the Moon. Although a finely comminuted component of Orientale ejecta may be present, it may be difficult or impossible to identify in the sample collection.

3.6 The formation and evolution of the Orientale basin

The Orientale impact occurred into a thick highland crust of dominantly anorthositic composition. Because this impact occurred fairly late in lunar history (about 3.8 Ga ago; Boyce, 1976; Wilhelms, 1987), the lunar lithosphere had already become rigid and subsequent post-basin endogenic modification was minimal. However, mare volcanism was active during this time: there are buried basaltic plains southeast of Orientale (Schultz and Spudis, 1979) and shallow magma chambers could have been present during the excavation of the basin and such magmatic ejecta could be entrained within the deposits of the basin (Schultz and Mendenhall, 1979; Ryder and Spudis, 1980).

Photogeologic mapping of pre-basin structures (Figure 3.10) helps to constrain the maximum size of the transient cavity of the Orientale basin. The preservation of older topography and structure within the Cordillera scarp (including parts of the outer Rook ring) suggests that the transient cavity could not have been larger than 620 km in diameter (the diameter of the outer Rook ring). In addition, Schultz and Spudis (1978) identified a radial structure that transected the inner Rook ring (Figure 3.10), but this feature is somewhat uncertain. The preservation of an extensive pyroclastic vent system between the Cordillera and outer Rook rings (Schultz and Spudis, 1978) suggests that the Cordillera ring cannot mark the rim of the Orientale transient cavity; this vent system would have been destroyed during basin formation. Moreover, abundant craters that pre-date the basin are preserved between the Cordillera and outer Rook rings, buried by Montes Rook Formation (Head, 1974a; King and Scott, 1978; Scott et al., 1977). The innermost limit of transient cavity is difficult to constrain, but the cavity is probably no smaller than the size assumed in models that equate the inner rings of basins with the original crater (Table 1.1), in the case of Orientale, about 500 km in diameter.

Results of remote-sensing studies suggest that only anorthosite-rich, upper levels of the crust were excavated during basin formation. In the Orientale region, this may have been as deep as 60 km (Bills and Ferrari, 1976), supporting models that indicate the effective depth of excavation of large impacts is relatively shallow. For a transient cavity of 500 to 620 km, excavation from depths shallower than 60 km suggests depth–diameter ratios of about 1:10 for the crater of excavation of large, basin-sized impacts. These results are consistent with the basin cratering model proposed by Croft (1981a), who adapted a simple hydrodynamic cratering flow model for terrestrial explosion craters developed by Maxwell (1977). Croft predicts that basin impacts will excavate to depths of approximately one-tenth the diameter of the transient cavity. For Orientale, this model suggests depths of excavation from 50 to 62 km, in good agreement with the values deduced from the orbital geochemical data and the crustal composition model of Ryder and Wood (1977). In the model of Croft (1981a), the actual transient cavity of the basin extends deep into the Moon, with a depth–diameter ratio approaching 1:3 (geometrically similar to that of small, simple craters; see sec. 2.1), but most of this material is simply displaced and rotated within the transient cavity, not ejected from the crater. The material that actually leaves the transient cavity must follow the flow streamlines that occur above the hinge streamline and thus, effective removal of material occurs from the much shallower, gravity-controlled *excavation crater*, in this case, from depths about one-tenth the diameter of the transient crater.

This model for the Orientale transient cavity has some implications for the origin of the basin rings. Photogeologic evidence of scarp morphology and truncated, older craters suggests that at least the Cordillera ring is of fault origin; an initially small transient crater further suggests that the outer rings (greater in size than the transient crater) are formed in response to crustal adjustments to the basin impact. This is the model advocated for outer ring formation by megaterrace formation in response to the excavation of large quantities of crustal materials (Head, 1974a; Dence, 1977a; McCauley, 1977; Scott *et al.*, 1977). The origin of the inner basin rings remains unclear; both the central uplift model (e.g., Howard *et al.*, 1974) and the oscillating peak model (e.g., Murray, 1980) are consistent with the observed geological relations.

Ejecta from the Orientale basin impact are composed mainly of anorthositic rocks. There are no significant quantities of mafic material or KREEP; the Orientale impact was not large enough to excavate lower levels of the crust in this region of the Moon. Photogeologic evidence suggests that some quantity of impact melt is contained within the continuous ejecta, but a unique geochemical signature of this melt material cannot be determined from the present poor-quality remote-sensing data. The main melt sheet contained within the basin interior was not over-flown by the Apollo orbital geochemical instruments, but based on the regional ejecta composition and considerations of the cratering model suggested by study of terrestrial impact craters, the Orientale melt sheet probably has a composition similar to anorthositic norite, lacking in KREEP. This presumed composition is in strong contrast to impact melts found at the Imbrium and Serenitatis basins, both of which apparently sampled deeper crustal levels and produced LKFM melt rocks (Chapters 6 and 7). The mafic nature of these impact melts is a consequence of both bigger impact events, in the case of Imbrium, and the thinner crust found on the lunar near side at these basin sites (see Chapters 6 and 7).

Orientale basin ejecta appear to be homogeneous over large areas (Table 3.2) and there is no geochemical contrast between the continuous and discontinuous deposits. This relation suggests that the highlands of the western limb are fairly uniform in composition and the question of the proportions of local and primary ejecta in the continuous deposits of the basin cannot be answered at Orientale. This problem can be addressed at other basins where such geochemical contrast between ejecta and local materials is evident.

If the lunar sample collection includes Orientale ejecta, it remains unrecognized. The large size of a basin-forming impact implies long residence times (on the order of tens of minutes) for ejecta within the transient cavity prior to ejection and therefore increased likelihood for comminution of ejecta to a fine-grain size (Schultz, 1978; Schultz and Mendell, 1978). If such comminution has occurred, a "cryptic" component of basin ejecta may be present in all highland soils. This basin component would not be recognizable petrologically or geochemically and may not be identified until complete basin ejecta blankets are studied in detail on the Moon.

For many years, the Orientale basin served as the type example of the lunar multi-ring basin and as a study area for basin formation mechanics and geology. However, I believe that the Orientale basin has several unique features which make its use as an archetype undesirable. The Orientale impact occurred into a thick highland crust, producing a unique ejecta composition and melt sheet chemistry. It is the youngest major basin on the Moon and formed after the lunar lithosphere largely had attained its present configuration (Solomon and Head, 1980), which argues against any substantial influence by endogenic processes during the modification stage of basin formation, in striking contrast to other older and more heavily modified basins. These differences in geological setting, age, and morphology of Orientale suggest that it should not be considered the archetype lunar basin, although study of its formational mechanics can lead to a better understanding of the excavation and modification (including ring creation) stages of basin formation.

4

An ancient basin: Nectaris

The Nectaris basin is located on the lunar near side (Figure 1.1), south of Mare Tranquillitatis and west of Mare Fecunditatis. An origin by impact for the Nectaris basin was advocated first by Baldwin (1949, 1963), who paid particular attention to the development of the Altai scarp, the southwestern topographic rim of the basin (Figure 4.1). The basin is relatively well preserved and served as a prototype multi-ring basin in the pioneering study of Hartmann and Kuiper (1962). More recent systematic studies of the Nectaris basin are those of Whitford-Stark (1981b), Wilhelms (1987), and Spudis *et al.* (1989). The Apollo 16 mission to the Descartes highlands in 1972 collected samples and orbital data that are directly applicable to comprehension of Nectaris regional geology. In this chapter, I describe the geology of the Nectaris basin and synthesize various data into a geological model for its origin and subsequent development.

4.1 Regional geology and setting

The Nectaris basin formed in typical crust of the near side highlands and interacted with two older basins. The average thickness of the crust in the region is about 70 km (Bills and Ferrari, 1976). The surrounding terrain consists of heavily cratered highlands, except where buried by later mare basalts. Nectaris deposits are well preserved to the south and west of the basin, but have been buried to the north and east by the lavas of Maria Tranquillitatis and Fecunditatis (Figure 4.1). In the area known as the "central lunar highlands", the western half of this region (bordering Mare Nubium) consists of large (100 km), overlapping impact craters, such as Ptolemaeus and Alphonsus. In contrast, the eastern half of the central highlands appear to be deficient in such large craters. A possible explanation for this deficiency is the burial of heavily cratered terrain by deposits of the Nectaris basin which have somehow lost their original surface texture and are thus unrecognized.

The topography of the Nectaris basin has been influenced by the presence of at least two nearly obliterated pre-Nectarian basins. The Tranquillitatis basin to the north intersects the trace of the absent outer ring of the Nectaris basin. To the east, the rim of the Fecunditatis basin is tangential to the main outer ring of Nectaris.

These older basins intersect northeast of Nectaris and may be responsible for the production of a Nectaris tangential structure that disrupts the continuity of the main rim (the Altai ring; Figure 4.1). Moreover, the inner basin rings of Nectaris display considerable irregularity, which may have resulted from the influence of the outer rings of these two pre-existing structures.

Pre-Nectarian craters are recognized beneath the outer basin deposits, but not within the basin, in contrast to Orientale (Chapter 3). This lack of pre-basin craters may be caused in part by the greater age and hence advanced state of degradation of the Nectaris basin.

Figure 4.1 Basin-centered view of the Nectaris basin. The Altai scarp (A) marks the topographic rim of the basin. Rectified telescopic view from Lunar and Planetary Laboratory, University of Arizona.

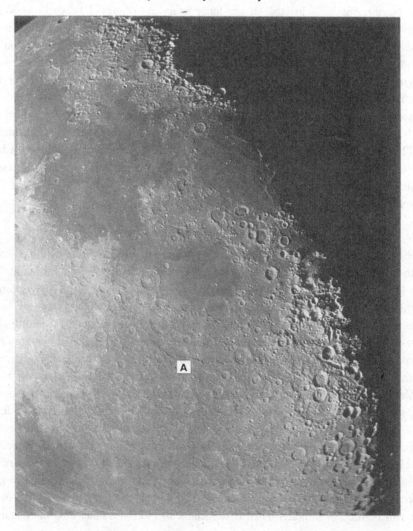

4.2 Nectaris morphology and geological units

A sketch map (Figure 4.2) summarizes the distribution of geological units in the Nectaris basin area. Nectaris displays several morphological facies that indicate a

Figure 4.2 Geological sketch map of the Nectaris basin. Deposits are largely preserved, although modified by superposition of plains and secondaries from the Imbrium basin. Base is a Lambert equal-area projection centered on 16° S, 34° E (Nectaris basin center). After Spudis *et al.* (1989).

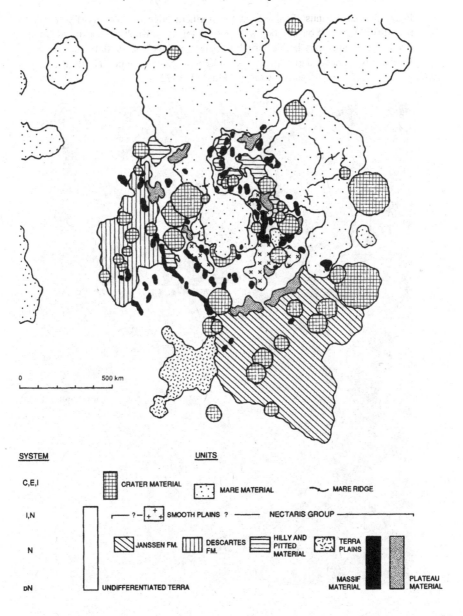

variety of depositional environments and unit provenance. The pattern of distribution of units of similar scale and morphology suggests that Nectaris is comparable in many ways to better exposed, less degraded basins, such as Orientale. However, because the Nectaris basin is old (hence degraded) and partly mare-flooded, complete characterization of all basin units is difficult. I use these incomplete unit exposures to infer the original morphologic properties of the Nectaris basin, tempering such inferences with the realization that Nectaris displays several unusual asymmetries that may invalidate direct comparison to the geological units of other basins.

Figure 4.3 Censorinus highlands at the northern edge of Nectaris. Platform massif (P) marks the basin topographic rim in this area. Knobby deposits (K) in intercrater regions may be Nectaris basin ejecta (Wilhelms, 1980a). Numerous Imbrium basin secondaries (S) mask underlying Nectaris units. North at top, framelet width 12 km. LO IV-72 H3.

Although most of the basin interior is mare-flooded, there are several exposures of basin floor materials, such as an exposure of smooth highland plains just south of Mare Nectaris (Figure 4.1). These smooth plains occur primarily between the inner and intermediate rings and consist of moderately high-albedo, undulating plains. The Nectaris smooth plains occur in a position similar to the fissured facies of the Maunder Formation of the Orientale basin and are here interpreted to be composed, at least in part, of melt deposits created during the Nectaris impact. These plains may be partly buried at the surface by Cayley material emplaced during formation of the Imbrium basin.

In intercrater regions northeast of Mare Nectaris, near the crater Capella, a knobby-like deposit (Figure 4.3) occurs sparsely, mainly in the vicinity of the intermediate ring of the basin. It has been interpreted by Wilhelms (1980a) as the Nectaris analog to the Montes Rook Formation of the Orientale basin. Alternatively, this unit could be a distal facies of Imbrium basin ejecta, as has been proposed for the Descartes material (Hodges and Muehlberger, 1981). No other exposure of this knobby material has been recognized in or near the Nectaris basin.

The continuous deposits associated with the Nectaris basin have been formally named the Janssen Formation (Stuart-Alexander, 1971; Stuart-Alexander and Wilhelms, 1975). The Janssen Formation (Figure 4.4) is exposed primarily to the southeast of the Nectaris basin and consists of hummocky to radially lineated deposits that extend nearly a basin diameter (850 km) from the basin rim. Many pre-Nectarian craters are visible in this region, suggesting that the Janssen Formation is less than one kilometer thick on average. The Janssen Formation grades radially outward into isolated basin secondaries and chains; as at Orientale, some secondary chains occur within the thick continuous ejecta sequence (e.g., Vallis Rheita; Figure 4.4b). The Janssen Formation is not recognized to the west of the basin; if it was ever there, the effects of Imbrium basin secondaries and distal continuous deposits probably have obliterated any radial surface texture.

Nectaris displays several unusual morphologic features outside of the basin. Topographically anomalous plateau areas (Figure 4.1, 4.3) are here named *platform massifs*. The Kant platform massif is rectilinear in plan and has an average elevation above the surrounding terrain of about 2 km. This platform is probably not solely of depositional origin; its outline suggests structural control. Several other platform massifs are evident in the Censorinus highlands (Figure 4.3), where they are associated with the tangential structure that occurs at the intersection of the Tranquillitatis and Fecunditatis basins (Figure 4.2). Platform massifs may be related to late-stage adjustments of the crust to the impact-excavated cavity (Schultz, 1976a, 1979; Spudis *et al.*, 1989); they will be discussed in the next chapter.

One of the most enigmatic geological units in the Nectaris region is the Descartes Mountains (Figure 4.5; Milton, 1972; Hodges, 1972). This unit consists of furrowed, wormy-textured materials that appear to overlie the uplifted Kant

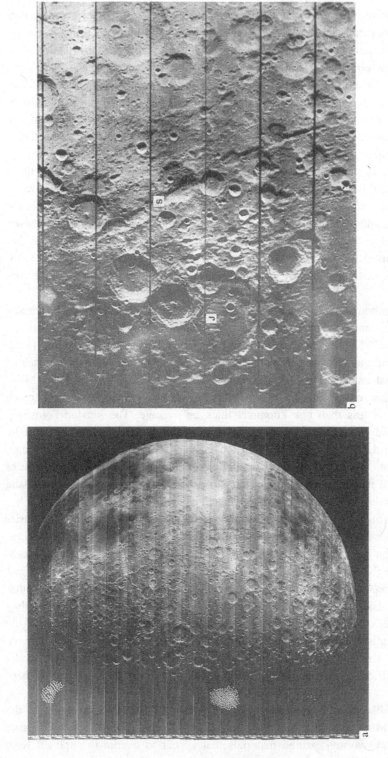

Figure 4.4 The Janssen Formation of the Nectaris basin. (a) Regional view, showing lineations and chain secondaries. Unit extends as far as 800 km from basin rim. Kant Plateau platform massif (K) also visible on northwest rim of basin. LO IV-83 M. (b) Detail view of the Janssen Fm. Most primary texture destroyed by impact erosion; chain secondary (S) still visible. North at top, large crater Janssen (J) is 190 km in diameter. Portion of LO IV-52M.

plateau platform massif (Figure 4.4a). It has been interpreted both as primary Nectaris ejecta (Wilhelms, 1972; Head, 1974b) and as Imbrium primary ejecta deposited as a deceleration lobe, analogous to those seen in the Hevelius Formation of the Orientale basin (Hodges and Muehlberger, 1981). If the Descartes material is primary ejecta from Nectaris, it is difficult to understand why similar appearing materials do not occur elsewhere around the basin. A variation of the Nectaris ejecta hypothesis suggests that this material has been sculpted by the arrival of Imbrium basin ejecta, producing secondary texture on top of the original Nectaris morphology. This type of "hilly and furrowed" material has been recognized in association with other basins such as Humorum (Wilhelms and McCauley, 1971;

Figure 4.5 The hummocky Descartes Formation (D) in the vicinity of the Apollo 16 landing site (A). Massif (M) is part of an exterior ring of the basin. North at top, framelet width 12 km. LO IV-89 H.

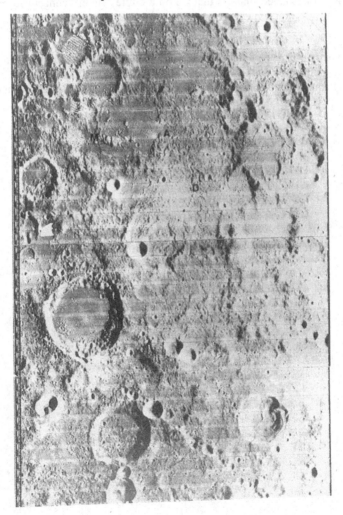

Howard *et al.*, 1974) as a type of basin ejecta facies (Wilhelms, 1980a). Thus, the Descartes may represent Nectaris ejecta whose morphology is a result of unusual composition, physical state of ejecta, energy environment during the deposition stage, or some combination of these factors.

The Nectaris basin displays five rings in various states of preservation (Figure 4.6; Table 2.1). A questionable innermost ring (240 km in diameter) is buried and represented only by a concentric system of ridges located within the confines of Mare Nectaris. Circumscribing this ring is a shelf-like massif ring (400 km in diameter), made up by the Montes Pyrenaeus in the eastern half of the basin. Most of the mare basalts are concentrated within this ring, which also marks the outer boundary of the positive gravity anomaly (mascon) of the basin (Sjogren *et al.*, 1974). The next larger Nectaris ring (620 km in diameter) is less well defined and passes through the craters Catharina in the west and Columbo in the east. This basin ring is composed of isolated massifs that lie on a circle crudely concentric to the main

Figure 4.6 Ring map of the Nectaris basin. Main rim (long dashes) is Altai ring (860 km diameter); other rings are 240, 400, 620, and 1320 km diameter. Base is a Lambert equal-area projection centered at 16° S, 34° E (Nectaris basin center). From Spudis *et al.* (1989).

basin rim. This ring may be analogous to the outer Rook ring of Orientale because the knobby deposits of Nectaris described by Wilhelms (1972, 1980a) appear to be concentrated near it.

The main rim of Nectaris (860 km in diameter) is well preserved in the south but appears to be discontinuous to absent in the northern half of the basin area. This ring is best exposed along the Altai scarp southwest of the Nectaris basin (Figure 4.1) and is here named the Altai ring. The Altai ring displays simple scarp-like morphology (Figure 4.7) and appears to be analogous to the Cordillera scarp of Orientale. Outside of the Altai scarp, textured exterior ejecta are poorly preserved and in some cases, obliterated, yet the basin ring is very sharply defined. One explanation for this relation is that this scarp has been rejuvenated during extended periods of crustal adjustment after the modification stage of basin formation (Hartmann and Wood, 1971; Schultz, 1976a). In this view, Nectaris would be analogous to those lunar craters displaying evidence for internal modification in the form of floor fractures (Pike, 1971; Schultz, 1976a, 1976b).

The extended period of scarp development would enable the original texture of the Janssen Formation to be obliterated by normal processes of degradation (possibly accentuated by the arrival of Imbrium materials) while preserving the pristine morphology of the outer basin scarp. Baldwin (1974, 1981) used the Altai scarp as *prima facie* evidence for his tsunami model for basin ring origin, citing the apparent contemporaneous ages for highlands on both sides of the scarp, determined by crater counts, as proof that the Altai scarp was created at the moment of Nectaris basin formation. However, Baldwin counted all highland craters in this region, many of them Imbrium secondaries (Wilhelms, 1980a). Therefore, these crater counts simply reflect the uniform age of the surrounding Nectaris highlands. Moreover, if the Altai scarp were formed by rejuvenation, apparent crater densities on both sides of the scarp would still be the same because the only activity is fault movement along the scarp.

Outside the Altai ring, a large (1320 km) but subtle exterior ring (Figure 4.6) is recognized (Hartmann and Kuiper, 1962; Wilhelms and McCauley, 1971; Pike and Spudis, 1987). This ring underlies the Apollo 16 site (Figure 4.5) and is expressed as an anomalous massif just west of the landing point. This massif is mapped as the central peak of "unnamed crater B" in Head (1974b), but that crater, if real, is likely a cluster of large, overlapping basin secondaries (Spudis, 1984) and would not possess central peaks. The large ring external to the main rim of the Nectaris basin has its analog at Orientale in the 1300 km diameter "Rocca" ring (Table 2.1; Chapter 3) and several large rings exterior to the Imbrium basin (Chapter 7).

Interaction of the Nectaris basin rim with the pre-existing Tranquillitatis and Fecunditatis basins may have been responsible for the tangential structures and platform massifs in the northern half of the basin. The Censorinus platforms (Figure 4.3) in particular seem to result from the intersection of the three regional basins, an intersection that structurally disrupted the highlands crust in this area.

The formation of the Kant platform massif (Figure 4.4a) likewise may be related to post-Nectaris events, as this massif occurs at the intersection of three large outer rings of basins (Nectaris, Serenitatis, and Imbrium; Figure 2.7), and to interaction of the Nectaris rim with Tranquillitatis basin ring structure to the north. Accentuation of topography caused by the intersection of large-scale structures has been suggested as an explanation for the prominence of several highland areas near basins elsewhere on the Moon (e.g., Head, 1977b; Spudis and Head, 1977).

Figure 4.7 The Nectaris basin Altai scarp. Lack of distinctive basin textured deposits near scarp, in contrast to preservation and prominence of scarp itself, suggests that the main rim may display scarp rejuvenation (Schultz, 1976a). North at top, framelet width 12 km. LO IV-84 H1.

4.3 Remote sensing observations of Nectaris basin deposits

Estimates of the regional composition of Nectaris basin deposits come from both the Apollo orbital geochemical experiments (Adler and Trombka, 1977) and Earth-based spectral measurements (Spudis *et al.*, 1989). The Apollo instruments measured the chemistry of the highlands from northeast of the basin near Gutenberg to the eastern shore of Mare Nubium, across the central lunar highlands (Figure 4.8). Both gamma-ray and X-ray data are available for this region, resulting in a fairly complete characterization of the chemical composition of the basin deposits (Spudis and Hawke, 1981).

The composition of the highlands across the Nectaris basin is diverse. The region to the northeast of the Nectaris basin (Capella) displays generally Al-rich chemistry, but is more mafic than the areas surrounding the Orientale basin (Table 4.1). Moreover, Th levels in this region of the Moon are typically higher than those seen in isolated highland areas (Metzger *et al.*, 1977, 1981) and in conjunction with

Figure 4.8 Index map for Nectaris basin compositional data described in text. Areas outlined refer to regions of orbital chemical data (Table 4.1); letters refer to spectra presented in Figure 4.9. Mercator projection, from Spudis *et al.* (1989).

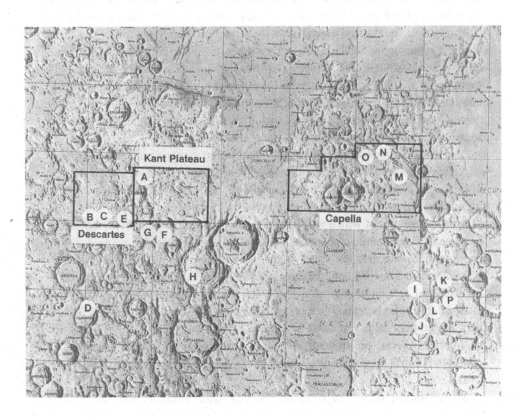

Table 4.1 *Chemical composition and mixing model results for deposits of the Nectaris basin*

Region	Al wt.%	Mg wt.%	Fe wt.%	Ti wt.%	Th ppm	FAN	ANGAB	NOR	TROC	LKFM	KREEP	11MB
Capella	14.6	5.1	5.1	1.1	1.2	—	74.7	—	—	25.3	—	—
Kant	15.4	4.9	3.7	0.5	0.6	29.3	—	33	12.3	—	9.9	15.5
Descartes	15.9	< 4.6	2.7	< 0.2	0.3	33.9	—	30.7	16.5	—	10.7	8.1
Descartes	"	"	"	"	"	21.2	47.4	—	—	27.7	—	3.7
Andel	12.2	4.6	5.0	0.7	2.2	22.8	—	54.2	4.2	—	10.5	8.4
Andel	"	"	"	"	"	9.2	63.6	—	—	—	14.7	12.5

Note: Regions outlined in Figure 4.8. Andel is region just west of map area (discontinuous Nectaris basin deposits).
Sources: Chemical data from La Jolla Consortium (1977), Davis (1980), and Metzger *et al.* (1981); mixing model results from Hawke and Spudis (1980) and Spudis and Hawke (1981).

the higher mafic content of the region, suggest that basin ejecta from crustal levels deeper than those from the Orientale basin are present.

The composition of the Kant plateau and Descartes region (Figure 4.8) differs from that of the eastern Nectaris highlands (Table 4.1). Kant has one of the highest Al/Si ratios of any geochemical province on the near side of the Moon (Andre and El-Baz, 1981; Spudis and Hawke, 1981). Moreover, deconvolution of Th abundances for this region indicates the materials here are extremely low in Th, comparable to some of the lowest Th levels seen in highlands areas of the lunar far side (Metzger *et al.*, 1981). Both of these measurements suggest compositions rich in anorthosite. If the materials of the Kant plateau and Descartes are related to Nectaris ejecta, as I believe they are, such material was probably derived from the uppermost crustal levels at the impact site of the Nectaris basin.

This Al-rich composition is limited to the Descartes region; increasing amounts of mafic components, as well as increasing Th, are seen in the highlands to the west of Descartes. Near the Apollo 16 landing site, the regional composition is similar to the eastern Nectaris highlands and representative of the compositions of the Cayley Formation, samples of which were returned from the Apollo 16 site. Moreover, the increase in Th and mafic component continues as the ground track is traced west of the Apollo 16 site near the crater Andel. An abrupt change in surface chemistry correlates with a distinct color boundary within the central highlands as seen on the multispectral vidicon mosaic of Soderblom (La Jolla Consortium, 1977, plate 7); I interpret this color boundary to have petrologic significance. The change in chemistry may reflect local variations in crustal composition resulting from deposition of Imbrium ejecta, pre-Imbrian volcanism, primary geochemical heterogeneity within the central highlands, or a combination of these factors (Spudis and Hawke, 1981).

Mixing model studies of these geochemical data, which include information from the X-ray spectrometer, were performed using both sets of end member compositions (subsec. 1.3.3). Good fits were obtained for all models; results are presented in Table 4.1. First, use of the mixed end members suggests that proportions of rock types are relatively constant for the Nectaris highlands, with the exception of the anomalous Descartes region described above. The ejecta of the Nectaris basin are best described as a mixture of anorthositic material and low-K Fra Mauro basalt in the approximate proportions of 3:1. In addition, minor amounts of mare basalt, which increase as the basalt deposits within Mare Nectaris are crossed, are required by the models. However, mare basalt within the basin ejecta seems likely, as the results indicate a mare component even within the highly anorthositic Descartes material (Table 4.1). This observation suggests that mare basalt was probably present at the impact site of the Nectaris basin. KREEP basalt becomes prominent west of the Apollo 16 site, replacing LKFM as the carrier of Th in this region. This relation may be caused either by the deposition of Imbrium ejecta or by pre-Imbrian KREEP volcanism. Pure anorthosite is present mainly in the Descartes region, where it is subordinate in quantity to anorthositic gabbro.

The second set of mixing model calculations used pristine end members (Table 4.1). These models suggest that the Nectaris highlands can best be described as a two-component mixture of anorthosite and norite as a roughly 1:1 mixture for the regions dominated by Nectaris basin deposits. Minor amounts of troctolite and KREEP are also present, as is a sizable portion of mare basalt, supporting the inference that mare basalt is included within the ejecta of the Nectaris basin. The KREEP component seen in these calculations probably represents the deeply derived component of Nectaris ejecta and is represented in the first set of calculations by LKFM. An abrupt change in the ratio of anorthosite to norite occurs west of the Apollo 16 site in the Andel region where norite becomes the dominant rock type, abundant over anorthosite by approximately 2:1. This change coincides with the highland color boundary described above and probably represents the margin of a major geochemical province on the Moon (Spudis and Hawke, 1981).

With the exception of the Descartes geochemical anomaly, the Nectaris basin ejecta deposits appear to be relatively uniform in petrologic composition and distinctly different from those of the Orientale basin. Both eastern and western Nectaris highlands may be described as mixtures of an anorthositic component in conjunction with LKFM in approximately a 3:1 ratio. Anorthosite is enriched in the Descartes region while KREEP is virtually absent, a consequence of the extremely low Th content of this region.

Spectral data of Nectaris basin units obtained from Earth confirm the inferences made from the orbital chemical data, with some interesting additions (Figure 4.9; Spudis *et al.*, 1989). Several outcrops of nearly pure anorthosite have been discovered in this region; these occur in the walls of Kant crater (A) and three places in the inner rings of the basin (F, J, and I in Figure 4.8; 4.9). Other spectra indicate the presence of mostly anorthositic norite rocks, with one deposit in the Censorinus highlands (Figure 4.3) of noritic composition. In general, the spectral data confirm the findings of the geochemical data; Nectaris basin deposits are dominantly noritic anorthosite to anorthositic norite (Spudis *et al.*, 1989).

From study of the remote-sensing data, the impact which formed the Nectaris basin produced an ejecta blanket of dominantly anorthositic materials. Results of the mixing models indicate the presence of substantial, but subordinate, quantities of a mafic and/or KREEP-rich component (norite, LKFM; Table 4.1). This component is almost totally absent in the Orientale ejecta blanket (Chapter 3) and because the two basins are comparable in size and probably excavated to similar depths, the presence of this mafic component may be a consequence of a relatively thinner crust at the impact site of the Nectaris basin. No ultramafic components (e.g., dunite) are seen in the Nectaris deposits in either the geochemical or spectral data,

For legend see next page.

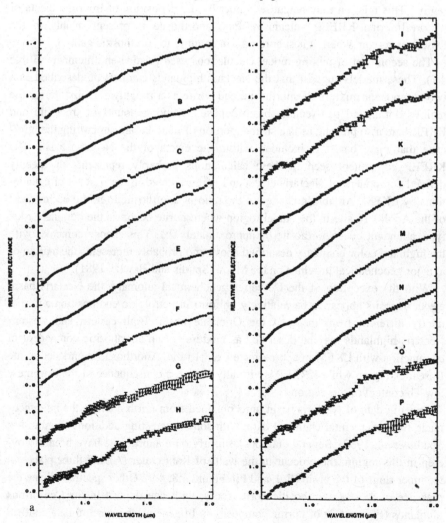

suggesting that no material from the lunar mantle was ejected within at least the outer basin deposits. These results suggest that the Nectaris basin excavated middle to lower crustal levels, but not below the crust of the Moon.

The presence in all mixing model solutions of a basalt component indicates that basalts make up part of the Nectaris ejecta sequence and that mare volcanism was probably active at the time of the Nectaris impact. This result is in accord with the global evidence that mare volcanism was active very early in lunar history (Schultz and Spudis, 1979; Ryder and Spudis, 1980). Moreover, the existence of mare volcanism at the time of the Nectaris impact provides a mechanism for the internal modification of basin topography in the manner described by Schultz (1979), in which extended periods of volcanic activity immediately after basin formation can cause such phenomena as scarp rejuvenation and the development of platform massifs.

Figure 4.9 Spectra of Nectaris basin geological units, discussed in text. (a) raw spectra; locations (letters) identified in Figure 4.8. (b) Selected spectra (continuum removed) for features described in text.

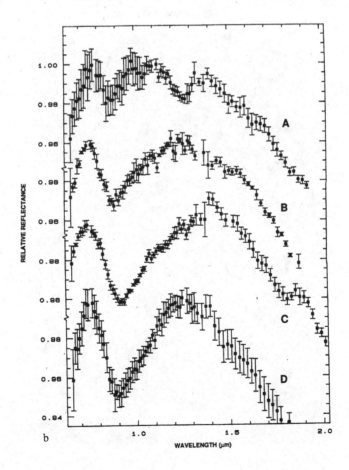

4.4 Apollo 16 site petrology – the Nectaris component

The Apollo 16 mission to the Descartes highlands returned a large collection of samples which, because of the proximity of the site to the basin, might be expected to contain Nectaris ejecta. In this section, several candidate samples of Nectaris basin ejecta from the Apollo 16 site are discussed within the regional context provided by the remote-sensing data. For detailed discussion of Apollo 16 site geology and petrology, see James and Hörz (1981), James (1981), Spudis (1984), and Stöffler *et al.* (1985).

Many Apollo 16 samples are impact melt rocks (Figure 4.10). These melts are relatively clast-free (Warner *et al.*, 1973a) and cluster into at least four compositional groups (McKinley *et al.*, 1984; Spudis, 1984), the most populous of which is of "VHA" (very high alumina) composition (Group 2, Figure 4.11). Low-K Fra Mauro basalt (Group 1) is the second most populous group; the other two groups correspond to high-Al varieties (Figure 4.11). Impact melts are widely distributed around the Apollo 16 site, although they appear to be concentrated within the

For caption see next page.

Cayley plains (G.J. Taylor *et al.*, 1973a; Warner *et al.*, 1973a; Ryder, 1981a). A survey of the Apollo 16 impact melts reveals a texturally heterogeneous collection (Figure 4.10), with changes in texture along clast boundaries as seen in other impact melt rocks from the lunar highlands (Simonds, 1975). The LKFM and VHA melt groups at the Apollo 16 site could be responsible for the LKFM component seen in the chemical mixing models of the orbital data (Table 4.1).

Current understanding of impact melt petrogenesis (Grieve *et al.*, 1977) suggests the Apollo 16 impact melts may have been generated by multiple impact events from large, local craters (Ryder, 1981a), but other considerations lend support to the hypothesis that many of them are impact melts from the Nectaris basin (Spudis, 1984). Composition of the Apollo 16 VHA impact melts is distinct from those associated with the Imbrium and Serenitatis impacts, yet the VHA composition (which appears to be an aluminous variety of LKFM) suggests derivation from deep within the lunar crust (Ryder and Wood, 1977; Spudis and Davis, 1986), by impacts much larger than the local craters evident near the Apollo 16 landing site. That these local craters cannot be responsible for the formation of VHA melt is shown by the regional composition of the Nectaris highlands (Table 4.1), which are both too aluminous and KREEP-poor to be the protolith for VHA impact melt

Figure 4.10 Petrographic textures of Apollo 16 VHA (group 2) impact melt, 61015; this group is interpreted here as Nectaris basin impact melt. (a) Mug shot of 61015 showing general texture of this "dimict" breccia; the VHA melt phase is dark. (b) Fine-grained intersertal texture (right) and very fine-grained subophitic texture (left) in melt phase. Cataclastic anorthosite at top right. Plane light, field of view 2.2 mm. (Courtesy O.B. James).

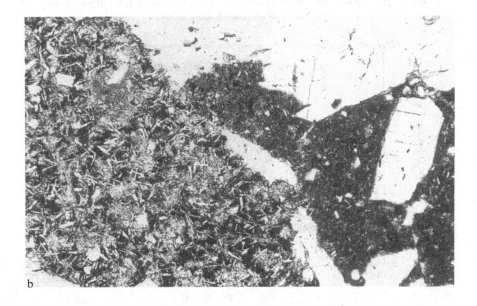

(Spudis, 1984). Moreover, if the VHA melts formed in local impacts, fragments of their protolith, which would be rich in KREEP basalt, should be present within the site deposits; none have been found.

These VHA melt samples are probably the best candidates for Nectaris ejecta at the landing site (Maurer *et al.*, 1978; Spudis, 1984) and may represent a melt component finely disseminated within a clastic sequence at the Apollo 16 site. No deeply derived clasts are observed within these melt rocks, as are seen in the Apollo 15 Imbrium basin impact melts (Ryder and Bower, 1977). This absence suggests that the VHA melts are derived from the upper levels of the "LKFM layer" within the lunar crust (Spudis and Davis, 1986). Such an upper level would also be the most likely region for the generation of a melt component that is ejected from the basin cavity.

Age data for Apollo 16 impact melts indicate that the VHA (Group 2 of McKinley *et al.*, 1984) melts formed at about 3.92 Ga ago. If these melt rocks are fragments of the Nectaris melt sheet, their age dates the Nectaris basin at 3.92 Ga old (Spudis, 1984). A different argument for this age of the Nectaris basin was given by James (1981) and the 3.92 Ga age was adopted by Wilhelms (1987) as the absolute age of the basin.

A group of melt rocks, named dimict breccias (James, 1981), have a VHA melt phase (McKinley *et al.*, 1984) and petrography similar to dike breccias found in ter-

Figure 4.11 Variation of Al and K in Apollo 16 impact melts. Proposed melt groups (1–3) of McKinley *et al.* (1984) shown. 67015,277 are melt clasts in fragmental breccia from Descartes material described by Marvin and Lindstrom (1983). Also shown is field for Apollo 17 poikilitic melt sheet, probably the impact melts from Serenitatis basin (see Chapter 6). Data from Ryder and Norman (1980); after Spudis (1984).

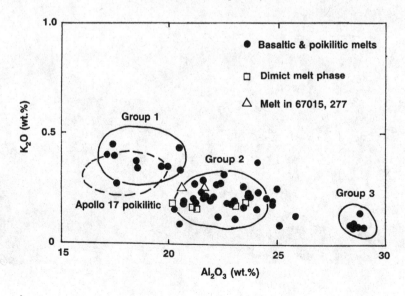

restrial impact craters. James (1981) interprets these rocks as having formed locally beneath a large crater as dike breccias, and being subsequently exhumed by other impacts. In contrast, I suggest that these rocks are impact melt from the Nectaris basin, formed where melt was forcibly injected into fluidized, unmelted debris surrounding the growing transient cavity of the basin (Spudis, 1984). If the dimict breccias are from a large crater instead of Nectaris, this crater cannot be local (because the melt phase is VHA; see above), but must occur uprange to the Imbrium basin and then would have been transported to the site by the Imbrium basin debris surge (Stöffler *et al.*, 1985).

Identification of the VHA melts with the Nectaris basin impact leaves the origin of the more mafic, LKFM group (Figure 4.11) unresolved. I have suggested that the Apollo 16 LKFM group may be melt from the Imbrium basin (Spudis, 1984), as they apparently formed somewhat later than the VHA group. However, Ryder and Spudis (1987) found that the melts identified from the Apollo 15 site, which probably contain Imbrium melt, do not resemble the Apollo 16 poikilitic LKFM group. I here suggest another possibility, that these rocks are impact melts from the Serenitatis basin; they fall into the compositional field defined by the Apollo 17 poikilitic melt sheet (Figure 4.11) and they have an age of about 3.87 Ga, the age inferred for the Serenitatis basin (Chapter 6). If they are from smaller craters near to the site (Stöffler *et al.*, 1985), again, these craters cannot be local and must occur uprange to the Imbrium basin, as the near-site materials do not have an LKFM composition.

Other impact melts returned from the Apollo 16 site (James, 1981; Ryder, 1981a; McKinley *et al.*, 1984) display considerable diversity in composition, but are generally more anorthositic than the LKFM and VHA melts (Group 3 of Figure 4.11). These melts probably represent of a number of impact events at the site (Ryder, 1981a). Their more anorthositic composition suggests generation in shallow level impacts, much smaller in scale than that of a basin. Head (1974b) believes that the regional geology of the Apollo 16 site is influenced by the presence of nearly obliterated pre-Imbrian craters of 75 and 150 km diameters (Figure 4.5). These craters are candidates for the generation of melt rocks of broadly anorthositic composition, but they cannot produce compositions similar to the LKFM and VHA melts. I conclude that the more aluminous impact melts described by Ryder (1981a) and McKinley *et al.* (1984) probably represent a spectrum of local impact events that are unrelated to the Nectaris basin ejecta at the Apollo 16 landing site.

Clastic breccias from the Apollo 16 site have been interpreted as Nectaris basin ejecta (James, 1981; Stöffler *et al.*, 1985). These rocks, called feldspathic fragmental breccias, make up most of the local ejecta from North Ray crater, which apparently excavated Descartes materials (Ulrich and Reed, 1981). Chemically, these rocks are very similar to the anorthositic components seen in abundance in the remote-sensing data (Table 4.1; Figure 4.9). I agree with Stöffler *et al.* (1985) that these rocks represent clastic ejecta from the Nectaris basin; I believe that these rocks are distinct from the melt phase of Nectaris ejecta, which are represented by the VHA impact melts. The inferred origins and geological relations of units at the Apollo 16 site are shown in Figure 4.12.

Samples returned by the Apollo 16 mission contain ejecta from the Nectaris basin. The best candidates for this material are the VHA basaltic impact melts for the melt ejecta and the feldspathic fragmental breccias for the clastic ejecta. While there is no assurance that either of these groups of samples is derived from the Nectaris basin, such a derivation is consistent with the regional context provided by the remote-sensing data.

4.5 The formation and evolution of the Nectaris basin

The Nectaris basin impact occurred into a typical section of near side highlands crust at a time when mare volcanism was active, probably about 3.92 Ga ago, based on the ages of the Apollo 16 VHA impact melts. At this time, the lunar lithosphere was still growing and relatively thin (Solomon and Head, 1980); the Nectaris impact may have penetrated the lithosphere, producing a mechanism to accentuate the structural modification of the basin ring system. The two pre-Nectarian basins Tranquillitatis and Fecunditatis destroyed the continuity of the outer Altai ring of

Figure 4.12 Geologic cross section of the Apollo 16 landing site. By the author, from Heiken *et al.* (1991).

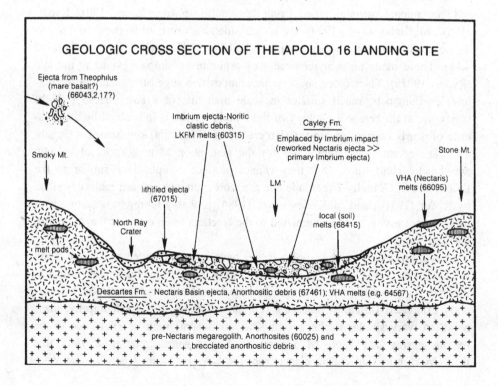

Nectaris and the intersection of these three basins produced the tangential structure of the Censorinus highlands. I postulate that both interaction with the older basin structures and penetration of the lithosphere by the basin cavity are responsible for the development of platform massifs surrounding Nectaris. This style of modification is not seen at the Orientale basin because of the thicker lithosphere in that region at the time of basin impact (Chapter 3).

The absence of recognizable pre-Nectaris topography within the basin appears to be a consequence of the basin's advanced state of degradation. Therefore, the size of the transient cavity of the Nectaris basin is uncertain. On the basis of the composition of Nectaris basin ejecta (Table 4.1), material below the crust was not excavated by this impact and the deepest ejecta were probably derived from middle crustal levels. Because the average crustal thickness in this region is about 70 km (Bills and Ferrari, 1976), the maximum depth of excavation in the Nectaris basin event was probably no deeper than about 50 km. The transient cavity of Nectaris was no larger than about 800 km (the diameter of the outer ring) and probably much smaller, as the Orientale results suggest that main basin rings do not represent the original crater rim (Chapter 3). The minimum value for the transient cavity diameter is about 350 km, the diameter of the innermost massif-like ring. In conjunction with results from the mixing model calculations, I suggest that the best estimate for the transient cavity of the Nectaris basin is slightly smaller than the diameter of the intermediate (Catharina) ring, about 500 to 600 km in diameter. As at Orientale, basin ejecta are derived from depths of about one-tenth the diameter of the transient cavity (Croft, 1981a). This estimate for the dimensions of the Nectaris basin transient cavity is not as certain as that for Orientale, where pre-basin topography narrowly constrains the dimensions of the cavity.

Immediately following the excavation phase of basin formation, the crust was modified to accommodate the unloading produced by ejection. In the case of the Nectaris event, this modification involved megaterrace formation to produce the Altai basin ring, possibly the Catharina ring as well, and internal sub-lithospheric flow accentuated the platform topography of the Kant plateau and the Censorinus highlands. Such a style of modification has been suggested as a common process in the formation of early lunar basins by Schultz *et al.* (1981) and presumes an initially small transient crater (Table 1.1), consistent with the model presented here. In this view, the Nectaris platform massifs result from foundering of crustal blocks aided by high sub-lithospheric heat content; such a thermal regime is suggested by ancient mare volcanism. Development of rigid blocks into mesa-like plateaux is seen in some cratering experiments where a rigid layer (lithosphere) overlies a more plastic (asthenosphere) layer (Greeley *et al.*, 1980). Topographic modification may have continued for extended lengths of time, possibly into the epoch when volcanism filled the inner basin, creating Mare Nectaris. Thus, the Altai scarp may have been rejuvenated some time after the time of the basin impact.

Ejecta from the Nectaris basin consist of a melange of middle to upper crustal products. The uppermost crust of ferroan anorthosite at the impact site appears to be concentrated in the Descartes materials. Ejecta derived from the lower portions of the excavation cavity are widely distributed and are intimately mixed with and subordinate in quantity to upper crustal debris over a wide region of the northern Nectaris ejecta blanket. Typical Nectaris ejecta consist of anorthositic material with lesser amounts of LKFM and norite and trace quantities of mare basalt. The approximately uniform concentration of mare basalt (about 10 percent for pure highlands regions) suggests that mare basalt was present both as surface deposits and, possibly, as near-surface magma reservoirs. The Apollo 16 data suggest that Nectaris ejecta were sampled and consist of a two-component mixture of impact melt (the VHA impact melt rocks) and clastic debris (feldspathic fragmental breccias). Such a mixture supports models that suggest ejected impact melt is an important component of basin ejecta. Impact melt flows segregating from textured basin ejecta are seen in the Orientale Hevelius Formation (Moore *et al.*, 1974; Wilhelms, 1987).

The relative contribution of material from the Imbrium basin impact to the deposits identified here with Nectaris basin are not known. Secondary cratering appears to be more important than blanketing in the central highlands, at least east of the Descartes region. The distinct change in surface composition near Andel crater (Table 4.1) may result from Imbrium ejecta, but it is also a possible relic of a pre-Imbrian geochemical province (Spudis and Hawke, 1981). The selective preservation of large craters in the western central highlands and their apparent absence near the Nectaris basin rim suggest that the ejecta from the Nectaris basin are present over most of the eastern central highlands and that the effects of subsequent deposition of Imbrium basin ejecta have been minimal. Secondary craters of Imbrium may have obliterated the surface texture of the Janssen Formation southwest of the basin and masked the pitted plains south of the Nectaris basin (Wilhelms, 1987). These plains may be old mare basalt flows, such as are inferred to have been present in the pre-Nectaris basin impact site, that were first buried by the deposition of Nectaris ejecta and then pitted by the impact of Imbrium basin secondaries.

The Nectaris basin displays several attributes that suggest an unusual style of modification. Platform topography in portions of the outer rim segment suggests that Nectaris is transitional in morphology between the Orientale-style basin and the Crisium-style basin (to be discussed in the next chapter). Changes in the modification style of lunar basins through time will be discussed in Chapter 8.

5

A modified basin: Crisium

The Crisium basin (Figure 5.1) was recognized as a multi-ring structure by Baldwin (1949, 1963) and Hartmann and Kuiper (1962), who were struck by its remarkable elliptical appearance. Although similar to other basins in the morphologic elements of ring massifs, ejecta, and secondary craters, Crisium displays several features that suggest it may have undergone a distinctly different style of post-impact modification. I will describe the regional and basin geology of the Crisium area and address the nature and causes of these morphological differences.

5.1 Regional geological setting

The Crisium basin (Figure 1.1) is on the eastern edge of the near side of the Moon, north of Mare Fecunditatis and southeast of Mare Serenitatis. The basin appears to have formed within a zone of typical highlands crust and mare volcanism was active in this region prior to the basin impact (Schultz and Spudis, 1979, 1983). The average thickness of the crust here is about 60 km (Bills and Ferrari, 1976). The interior of the Crisium basin is completely mare-flooded, which has obscured the relations of basin materials; this obscuration has resulted in controversy regarding the true topographic rim of the basin (Howard *et al.*, 1974; Wilhelms, 1980b, 1987; Croft, 1981b), as discussed below.

The Crisium basin (Figure 5.1) appears to have had minimal interaction with older basin structures. The nearest pre-Crisium basin is the Fecunditatis basin, whose center is located approximately 700 km to the south of Crisium. This basin is tangential to the outermost Crisium ring of about 1000 km diameter mapped by Wilhelms and McCauley (1971) and Fecunditatis effects on the generation of Crisium topography have probably been relatively minor. Head *et al.* (1978) postulate that the missing segment or notch evident in the eastern rim of Mare Crisium may be caused by the interaction of Crisium with a pre-existing basin that was nearly destroyed by the Crisium impact. Wichman and Schultz (1992) suggest that the notch in this part of the basin rim, and the general elongation of the Crisium rim, is a consequence of a highly oblique impact, a hypothesis I will discuss below.

Alternatively, this notch may be caused by modification of basin topography after the Crisium impact due to local variations in crustal or lithospheric thickness.

5.2 Crisium morphology and geological units

Although the Crisium basin is of Nectarian age (Wilhelms, 1987) and thus not pristine, the exterior and rim deposits of the basin are fairly well exposed in some sectors. In this section, I describe the geology of the Crisium basin (Figure 5.2) and infer possible geological processes that have produced its style of basin modifica-

Figure 5.1 Basin-centered photograph of the Crisium basin. The rim of massifs bounding the mare is about 540 km in diameter; re-entrant on east side of basin gives an elliptical appearance to the basin. North at top; horizontal field of view about 900 km. Lunar and Planetary Laboratory photograph, University of Arizona.

tion. Because mare flooding covers key relations within the basin interior, the conclusions I draw here from photogeology require more support from orbital geochemical data and sample information than do those for other basins discussed in this book.

Figure 5.2 Geological sketch map of Crisium basin deposits, after Wilhelms and McCauley (1971), Wilhelms and El-Baz (1977), Wilhelms (1987), and new mapping by author. Base is a Lambert equal-area projection of the eastern limb of the Moon (Rükl, 1972).

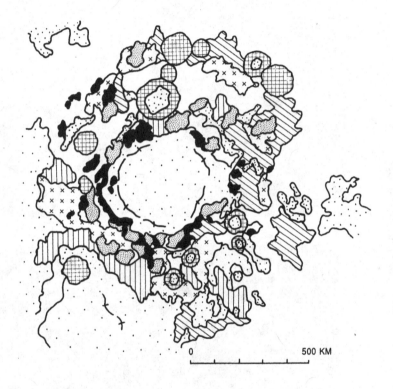

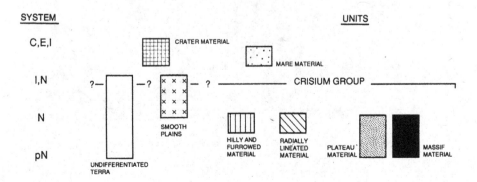

For legend see next page.

The interior of the Crisium basin is covered by mare basalt. Therefore, no inferences can be made regarding basin units within the structure that may be analogous to those in relatively unflooded basins, such as Orientale (Chapter 3). The bounding rim of the mare fill consists of well preserved massifs (Figures 5.2, 5.3) that have superficial resemblance to massif rings associated with basins such as Serenitatis, where extensive mare flooding has obscured geological relations. This innermost massif ring is actually the inner portion of a complex topographic rim system that may not be directly analogous to any other basin rim system.

The Crisium rim has an unusual step and platform topography. Polygonal, mesa-like islands contain undulating to smooth deposits. In places, these deposits resemble light plains or hummocky basin ejecta (Figure 5.3). The hummocky deposits may also overlie the platform massifs. Platform massifs are much more prominent and abundant than those described previously around the Nectaris basin (Chapter 4), and appear to make up most of the highlands that surround Mare Crisium for about a basin radius (250 km). These platforms are morphologically dissimilar to any terra landform observed at Orientale, but do resemble the Caucasus Mountains of the Imbrium basin, described in Chapter 7. Schultz (1979) suggests these platform massifs are produced by endogenic modification of a transient cavity that was originally much smaller than the present basin.

Highlands on the southern border of the Crisium basin display several morphologies that render interpretation of basin geology difficult (Figure 5.3). Large exposures of "hilly and furrowed material" (Wilhelms and McCauley, 1971) occur in this region and are somewhat similar in morphology to the Descartes materials of the Nectaris basin. Hilly and furrowed material was commonly interpreted to be of volcanic origin in the pre-Apollo-16 literature (Trask and McCauley, 1972; Schultz, 1976a) and units of similar appearance crop out in the vicinity of the Humorum basin (the Vitello Formation; Spudis *et al.*, 1992), as well as the Descartes material of the Nectaris basin.

The common interpretation of hilly and furrowed material following the Apollo 16 mission is that it is a facies of basin ejecta. Although this explanation is reasonable for most of the observed relations, it is not totally satisfactory for the Crisium hilly materials. The rim of Cleomedes, a post-Crisium crater on the northern rim of the basin (Figure 5.3a), appears to be overlain by a deposit of this hilly material. If

Figure 5.3 Rim units of the Crisium basin. (a) Platform massifs (M) make up the main basin rim; interplatform areas covered by light plains deposits (P). Hilly, rough-textured materials (R) overlie both Crisium rim and crater Cleomedes (C). This unit probably post-dates Crisium basin: possibly it is ejecta from the younger Serenitatis basin. North at top; field of view about 600 km. LO IV-191 H3. (b) Close-up view of western rim of Crisium basin, showing mesa-like platform massif (M), ring trough (T), and lineated texture, probably Crisium ejecta (E). Crater Proclus (P) is 28 km diameter and probably formed from an oblique impact; dark halo craters are seen within its ray-excluded zone (Figure 5.4). North at top; Apollo metric AS17-M-293.

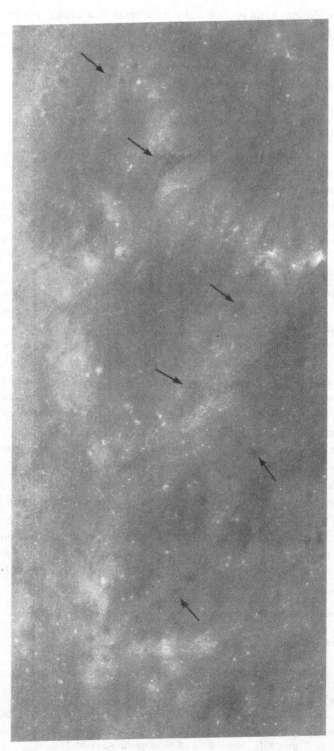

Figure 5.4 Dark-halo craters (arrows) within the interplatform smooth plains of Crisium, near the crater Proclus (see Figure 5.3b). These small impact craters have excavated mare basalts that flooded Crisium interplatform areas and ring troughs. Width of scene about 20 km. Part of AS15-9234 (pan).

this relation is correct, the hilly material cannot be a facies of Crisium basin ejecta because it overlies a crater that post-dates the Crisium basin. Such a stratigraphic relation is similar to that of the Sculptured Hills material of the Serenitatis basin, to be described in the next chapter.

The Crisium rim is surrounded by hummocky to knobby deposits that overlie the basin platform structure. These units are probably primary ejecta of the Crisium basin, discontinuously covered by mantling deposits from later basins (e.g., Serenitatis, Imbrium). The hummocky units appear similar to continuous deposits surrounding other lunar basins, yet radially textured ejecta comparable to the Hevelius Formation of the Orientale basin (Chapter 3) are preserved poorly. Undulating smooth plains deposits, resembling the Cayley plains (Figures 5.2, 5.3), occur within interplatform areas. Some of these deposits, particularly near the crater Proclus, display dark-halo impact craters (Lucchitta, 1972; Figure 5.4), evidence that they are, at least in part, underlain by old mare deposits (Schultz and Spudis, 1979). Because the interplatform areas are partly flooded by mare basalt, the light plains are not solely of Crisium basin origin; mare plains within topographic lows of the Crisium rim would have to post-date the basin. Because the light plains overlie the ancient mare deposits, they are at least partly of non-Crisium basin origin; the Serenitatis and Imbrium basins are likely sources for some of these materials.

Distal portions of Crisium basin ejecta are difficult to recognize in most regions (Wilhelms, 1976a, 1987; Wilhelms and El-Baz, 1977). Sparse patches of lineated terrain occur north of the basin, terminating near the crater Messala (Figure 5.2), and probable basin secondaries are found near the crater Zeno (Figure 5.5). Wilhelms (1976a) postulates that the elongate crater Rheita P, superposed on the Janssen Formation of the Nectaris basin, may be a Crisium basin secondary crater. This interpretation, which I accept, implies that the Crisium basin post-dates the Nectaris basin.

5.3 Structural geology and rings of the Crisium basin

There is considerable controversy regarding the topographic rim of the Crisium basin, as previously mentioned. A subtle ring structure of about 1100 km diameter has been interpreted by Wilhelms (1980b), Croft (1981b), and Wichman and Schultz (1992) as the topographic rim of the basin. Other investigators prefer to interpret the scarp-like ring (740 km in diameter) just outside the massif ring that borders Mare Crisium as the main rim of the basin (Hartmann and Kuiper, 1962; Baldwin, 1963; Casella and Binder, 1972; Howard *et al.*, 1974).

Examination of the regional topography surrounding the Crisium basin (Figure 5.6) shows that the supposed "inner" rings of Crisium (i.e., the highlands that border Mare Crisium) collectively make up the abrupt edge of the mare basin; the surrounding highlands appear to be of relatively constant elevation. Highland elevations of the 740 km ring equal or exceed those of the ring bordering the mare

(Figure 5.6). I believe that the topographic rim of the Crisium basin is probably represented by the scarp–massif ring that occurs within these highlands (740 km in diameter); this assignment of the Crisium rim differs from the proposed main rim of Pike and Spudis (1987), who chose the 540 km diameter ring. If the outer 1100 km ring were the main topographic rim, units in the southern Crisium highlands might be expected to display morphological similarities with basin interior units, as observed at Orientale and Nectaris; no such analogous units are evident near Crisium. The controversy over the location of the basin rim is caused partly by the unusual modification style of Crisium that has made the example of Orientale of little use.

My reconstruction of the Crisium basin ring structure is shown in Figure 5.7. In this interpretation, the main topographic rim of the Crisium basin is represented by

Figure 5.5 Distal deposits of the Crisium basin. Cluster of craters superposed on rim of Zeno (Z) are probably Crisium basin secondary craters. Rough deposits (R) south of Messala (M) may be isolated patch of Crisium continuous deposits. North at top; picture width about 300 km. LO IV-177 H2.

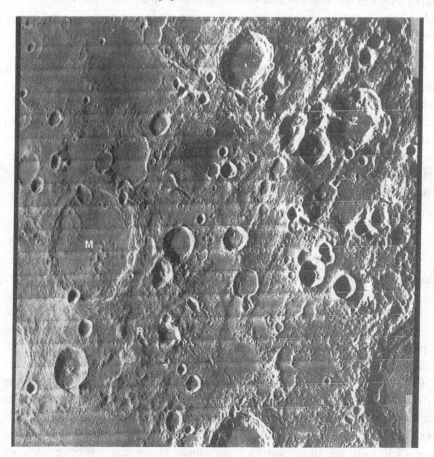

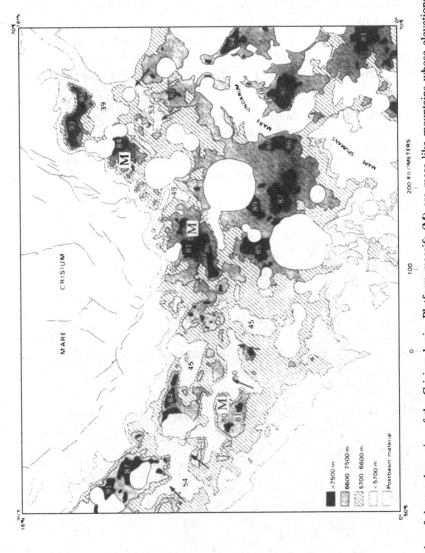

Figure 5.6 Topography of the southern rim of the Crisium basin. Platform massifs (M) are mesa-like mountains whose elevations correspond to the regional average elevation of the highlands in this area. Ring troughs (arrows) also evident; see Figure 5.8. Portion of LOC-3 topographic map of the Moon by U.S. Geological Survey; from Wilhelms (1987).

the prominent ring (740 km diameter) in the circum-Crisium highlands. The mare-bordering ring inside of the main rim (540 km diameter) resembles this rim morphologically, but displays about the same elevation or is slightly lower (Figure 5.6). Two large, exterior rings of 1080 and 1600 km diameter possess scarp-like morphology; they resemble the outer rings of the Nectaris and Imbrium basins (Table 2.1). One additional inner ring (360 km diameter) is expressed by the wrinkle ridge system of Mare Crisium (Figures 5.2, 5.3a, 5.7).

Among the most distinctive structural features of the Crisium basin are concentric troughs between the platform massifs that make up the basin rings. These troughs are evident both on low-sun Earth-based telescopic photographs (Figure

Figure 5.7 Ring interpretation for Crisium basin. Both the 540 km ring (2nd) and 740 km ring (3rd) have comparable average elevations (see Figure 5.6), but I here adopt the 740 km ring as the main basin rim (see text). Other rings indicated by short dash lines. Base is Lambert equal-area projection centered on 18° N, 59° E (center of Crisium).

5.8) and in the topographic data derived from orbiting spacecraft (Figure 5.6). The troughs appear to be structurally controlled, for they display a slightly polygonal outline; such an outline is in part caused by the coalescence of irregular platform massifs that make up elements of the ring troughs. Ring troughs are not found at any of the other lunar basins discussed here and they may reflect internal modification of basin morphology (Schultz, 1976a, 1979).

The east–west elongation of Mare Crisium is caused by a break in the continuity of the mare-bounding massif ring in the eastern portion of the basin (see above). This area has been proposed as the site of an older basin that disrupted the formation of the main topographic ring (Head *et al.*, 1978). If this hypothesis is correct, the pre-Crisium basin could be analogous to the pre-Orientale two-ring basin that intersects the main basin rim (see Chapter 3). Considering the unusual structural modification form displayed around the basin, an alternative idea is that this break is an internally generated feature, produced during modification of basin topogra-

Figure 5.8 Ring troughs of the Crisium basin. Troughs display *en echelon* offsets (arrows), suggesting they are not simple, inter-ring lows, but have developed by large-scale, structural modification of the entire region. North at top, field of view about 700 km. Lunar and Planetary Laboratory photograph, University of Arizona.

phy. Wichman and Schultz (1992), emphasizing the elliptical plan of the Crisium basin, argue that an oblique impact has produced an asymmetric distribution of ring elements. In my opinion, the arrangement of basin units is remarkably circular (Figures 5.2, 5.7); the elliptical plan of the basin is largely an illusion based on the distribution of mare deposits. While I do not reject the idea that Crisium may have formed during an oblique impact, I am more struck with the symmetries of Crisium than its asymmetries.

The structural modification style at Crisium may be caused by an anomalously thin lithosphere at the time of basin formation; the lithosphere could have been penetrated by the basin transient cavity. Mare basalt deposits that apparently post-date the basin within the interplatform regions of the basin exterior suggest that extensive volcanic activity immediately after the Crisium impact accompanied this extensive basin modification, as described on a theoretical basis by Schultz *et al.* (1981). Thus, the Crisium basin may be a scaled-up analog of internally modified craters such as Gassendi and Pitatus (Pike, 1971; Schultz, 1976b).

5.4 Composition of Crisium basin deposits

The Apollo 15 ground track (Figure 5.9) covers the southern and western rim of the Crisium basin and provides both gamma-ray and X-ray data for these highland regions. Post-Crisium volcanic flooding of the interplatform regions (Figure 5.4) could affect the composition of highlands in this area, but the outlines of regions

Figure 5.9 Index map of regions of orbital geochemical data given in Table 5.1. Units were defined by Bielefeld *et al.* (1978); unit D represents interplatform areas corresponding to old mare basalt flows covered by highland plains (Figure 5.2). After Bielefeld *et al.* (1978).

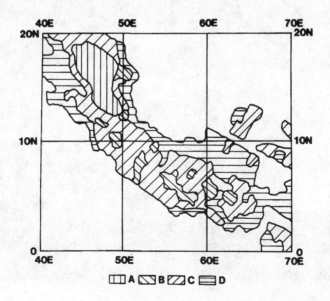

Table 5.1 *Geochemical data and mixing model results for deposits*
of the Crisium basin

Unit	Al wt.%	Mg wt.%	Fe wt.%	Ti wt.%	Th ppm	ANGAB	LKFM	12MB
A	12.6	4.8	4.9	0.3	1.8	75.5	22.7	1.8
B	12.9	4.9	4.8	0.3	1.7	78.8	21.2	—
C	12.5	4.8	4.8	0.3	1.7	75.1	22.3	2.6
D	11.2	5.3	7.1	0.6	1.7	61.4	19.9	18.7
Luna 20	12.5	5.8	5.7	0.3	1.3	66.5	27.9	5.6

Note: Units outlined in Figure 5.9.
Sources: From Bielefeld *et al.* (1978). Mixing model results from Hawke and Spudis (1979). Luna 20 soil composition from Papike et al. (1982).

selected for mixing model calculations were taken from the correlation study of Bielefeld *et al.* (1978) to minimize this effect and model only the ejecta deposits of the Crisium basin (Figure 5.9).

The regional composition of the southern rim of the Crisium basin is typical of near side highlands (Table 5.1). Al values are similar to those of the northeastern highlands around the Nectaris basin and both of these localities typify comparable geological environments. Moreover, extremely high Al values are associated with the Crisium platform massifs near the craters Apollonius and Taruntius X. These data suggest that anorthositic material dominates and mafic components are of lesser abundance. Unit "D" of Bielefeld *et al.* (1978) has the lowest albedo (0.125) of all units, generally lower Al/Si values, and correlates with the smooth interplatform regions, which I believe are underlain by mare volcanics (cf. Figures 5.2 and 5.9). Unit D exhibits dark halo craters in some regions, supporting this interpretation (Schultz and Spudis, 1979; Schonfeld, 1981).

Mixing model calculations were performed on the regional units mapped in Figure 5.9 (Hawke and Spudis, 1979). All four units produced very good fits with the mixed end members and also appear to be more or less constant in petrologic make up (Table 5.1). The three "pure" Crisium highlands regions (A, B, and C) are nearly identical in composition, being composed of anorthositic gabbro and low-K Fra Mauro basalt in the approximate proportions of 3:1. A minor mare basalt component of a few percent is seen in most units (except B), but the quantity of this component increases substantially in unit D (almost 19 percent). The models for Crisium basin ejecta are comparable to those obtained for Nectaris, in which anorthositic components make up the majority of basin deposits and LKFM is present in lower abundance. It is likely that both basins ejected material from comparable stratigraphic levels (not necessarily comparable depths) in the lunar crust.

From these results, several conclusions may be drawn regarding the nature of the basin impact site. It appears to have been a typical highlands crust in which the majority of ejecta are derived from an upper, anorthosite-rich layer; middle and lower

crustal levels are represented by a subordinate fraction of LKFM basalt. The petrographic nature of the LKFM component cannot be determined from the orbital data,
but it may be entrained within the continuous deposits of the Crisium basin as ejected
melt, as postulated for the ejecta of the Nectaris basin at the Apollo 16 landing site
(Spudis, 1984; see Chapter 4). There is no evidence for large amounts of mafic material or any ultramafic component in Crisium basin ejecta, indicating that excavation
was confined to middle and upper crustal levels. Mare basalts make up at least part of
the ejecta sequence, but because of post-basin mare flooding of the interplatform
regions, the actual amounts of basalt in the basin ejecta are uncertain. Mare basalt is
apparently in lower abundance than it is in the Nectaris basin deposits.

Ejecta from the Imbrium and Serenitatis basin impacts are not immediately
apparent. Both of these basins probably contributed a layer of discontinuous debris
to the Crisium region, but this ejecta probably would constitute less than 10 percent
of the observed sample and would not necessarily be distinct geochemically from
the Crisium deposits. Both the Imbrium and Serenitatis basins post-date the
Crisium basin and these features may be responsible in part for the deposition of
the Cayley-like smooth plains that occupy the interplatform regions and that fill
depressions in the ring troughs of the Crisium basin.

Figure 5.10 Geologic cross-section showing regional setting of the three Soviet
Luna sample-return missions. Luna 20 returned regolith developed on ejecta
from the Crisium basin. By the author, from Heiken *et al.* (1991).

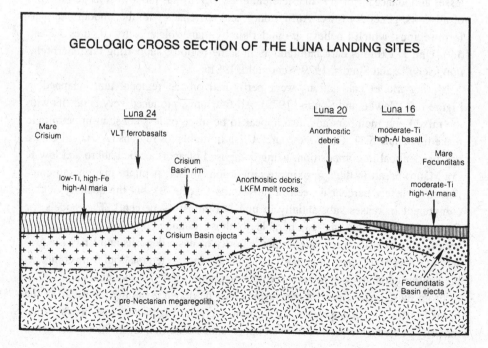

Table 5.2 *Chemical composition and modal data for Luna 20 glass particles*

	Anorthosite	ANGAB	LKFM	Mare basalt	KREEP
Ti wt.%	0.04	0.23	0.5	1.0	1.0
Al wt.%	18.6	13.8	10.6	6.9	7.9
Mg wt.%	0.48	4.8	7.2	4.0	3.6
Fe wt.%	0.52	4.5	7.3	12.3	7.6
Mode wt.%	8	65	18	9	0.4

Source: Data from Warner *et al.* (1972a), based on analysis of compositional clusters of impact glass compositions

5.5 Crisium ejecta: petrology of the Luna 20 site

The Soviet unmanned spacecraft Luna 20 landed on the southern rim deposits of the Crisium basin (Figure 5.10) near the crater Apollonius. It returned to Earth a 50 g sample of lunar soil within a shallow drill core that prevented the sampling of any large rocks. Thus, this regolith, the smallest sample return of any highland landing site on the Moon, may not be representative of the lunar crust in this region. However, the Luna 20 samples provide ground truth for the interpretation of orbital geochemical data, in particular, it may be used to calibrate the results obtained by the X-ray experiment to convert Al/Si ratios into Al concentrations (Bielefeld, 1977). As noted in the previous section, the Apollonius region displays enhanced concentrations of Al as seen from the X-ray data and therefore, the Crisium ejecta in this area may be atypical, compared with the basin deposits as a whole.

Many particles within the soil of this region consist of impact-derived glasses. These particles may be analyzed chemically by electron microprobe (Warner *et al.*, 1972a); such an exercise indicates clusters of compositions that correspond to major lunar rock types. A summary of the major chemical compositions observed in the Luna 20 samples is presented in Table 5.2.

The compositions of glass and small lithic fragments from the Luna 20 soil suggest that anorthosites and anorthositic gabbros are the most abundant component of the soil, making up about 75 percent of the sample (Warner *et al.*, 1972a; Prinz *et al.*, 1973; G.J. Taylor *et al.*, 1973b). The relative amount of this material is in remarkable agreement with the results inferred from the mixing models of orbital chemical data (cf. Tables 5.1 and 5.2). No pristine anorthosites were found, probably a consequence of the small sample return. Lithic fragments with slightly more mafic composition are also found in the Luna 20 soil and appear to resemble anorthositic troctolite (Cameron *et al.*, 1973; Prinz *et al.*, 1973); these fragments may represent mafic portions of larger anorthosites.

The next most abundant composition is equivalent to low-K Fra Mauro basalt, which contributes about 18 percent to the total Luna 20 soil (Warner *et al.*, 1972a). These glasses are similar in composition to and may be derived from lithic debris in the samples described as high alumina basalt by Prinz *et al.* (1973). LKFM fragments from the Luna 20 soils are similar in bulk chemistry to the Serenitatis basin Apollo 17 impact melts (Chapter 6). This soil component is probably represented in the orbital mixing model calculations by the LKFM component (Table 5.1); the abundance of LKFM as seen from orbit is consistent with this interpretation.

In addition to the highland rock types in the Luna 20 samples, a minor component of mare basalt (impact glasses and pyroxene mineral fragments) has been observed (Warner *et al.*, 1972a; Cameron *et al.*, 1973). Mare-type glasses make up about 9 percent of the soil; the estimates of mare basalt abundance from the mixing model calculations range from 2 to 18 percent, again in reasonable agreement. These mare basalt fragments may be derived from the post-basin Mare Crisium and Mare Fecunditatis flows, delivered to the site by impact craters (Heiken and McEwen, 1972), but the evidence for ancient, basin-related mare deposits, both as ejecta and post-Crisium (but pre-Imbrium) flows, suggests that at least some of these fragments may be part of the ejecta from Crisium basin at the Luna 20 site.

Among the regolith particles returned by the Luna 20 mission are fragments of fine-grained, crystalline rocks (G.J. Taylor *et al.*, 1973b; Swindle *et al.*, 1991). Many of these rocks are impact melt fragments of aluminous highlands composi-

Figure 5.11 Sc–Sm chemical composition of Luna 20 rocks. The dispersed nature of this plot suggests that samples are derived from multiple events, some of which may be related to basin-forming impacts. Envelopes show the range of compositions for the Apollo 17 aphanites and poikilitic impact melts (Serenitatis basin melts; see Chapter 6) and the Apollo 16 VHA basaltic impact melts (Nectaris basin melts; see Chapter 4). Solid symbols are rocks known to be impact melts; the petrography of open symbols is not known, but many could be melt fragments. Data from Swindle *et al.*, 1991 (squares), Korotev and Haskin, 1988 (circles), Smith *et al.*, 1983 (diamonds), and Laul and Schmitt, 1973 (triangles). After Swindle *et al.* (1991).

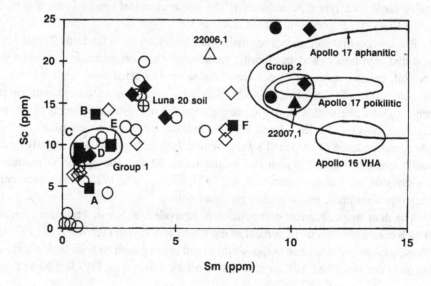

tion, containing significantly lower KREEP than the melt rocks of other highlands sites (evident from lower Sm content; Figure 5.11). The melt rocks display a variety of compositions, but appear to be significantly lower in Sc and Sm than the mafic, basin-related melt rocks at the Apollo highland sites. However, some Luna 20 melt rocks have compositions comparable to LKFM and VHA basaltic impact melts (e.g., Group 2 and isolated melt samples (including "F") in Figure 5.11). While the aluminous melt rocks can be explained as fragments of the melt sheets of small, local craters near the Luna 20 site, these more mafic samples cannot be derived from such craters, as they are richer in both Sc and Sm than the local substrate (note position of "soil" in Figure 5.11).

The cluster labeled "Group 1" on Figure 5.11 is clearly of highlands derivation (Al_2O_3 about 28%, based on an FeO content of 4.6%), forms a tight cluster, and is well populated; it very likely represents an impact melt sheet of a small, local crater. Other samples, being of a more mafic composition than the soil and less abundant, might represent basin impact melt. On the basis of similarity to Apollo 17 impact melt and a contemporaneous Serenitatis basin age for one of the samples (22007,1, dated at 3.87 Ga by Podosek *et al.*, 1973), Swindle *et al.* (1991) concluded that Group 2 melts from Luna 20 probably come from the Serenitatis basin and that a single particle ("F" in Figure 5.11) is a fragment of the impact melt sheet of the Crisium basin. This sample (22023,3,F) from Luna 20 was dated at 3.895 ± 0.017 Ga (Swindle *et al.*, 1991), an age consistent with such an interpretation.

Fragment "F" and others of similar composition probably represent ejected impact melt from the Crisium basin that is included within clastic ejecta of anorthositic composition at the Luna 20 landing site. Such a geological relation is similar to the occurrence of the VHA melt rocks included within a sequence of clastic debris at the Apollo 16 site; this deposit was probably derived from the Nectaris basin impact (Spudis, 1984; sec. 4.4). On the basis of such an analogy, much of the Luna 20 sample is derived from primary ejecta of the Crisium basin; material from later basin impacts (e.g., Serenitatis) may also be present in lower abundance. This Crisium basin material consists largely of anorthositic debris, some of it re-melted by local impacts (e.g., group 1 melts; Figure 5.11).

The composition of the Luna 20 samples agrees with the petrologic composition of Crisium basin deposits inferred from mixing models of orbital data. This agreement implies that primary ejecta from the Crisium basin were sampled at the Luna 20 site (Figure 5.10) and that the ground truth provided by this sample gives confidence in the regional interpretations derived from the remote-sensing data.

5.6 The formation and evolution of the Crisium basin

The Crisium target appears to have been similar chemically and petrologically to the impact site of the Nectaris basin described in the previous chapter. The highlands crust in this region was probably about 60 to 70 km thick at the time of basin impact and mare volcanism was apparently active. The thermal state of the crust in this region of the Moon was quite different than it was for the impact sites of the Orientale and Nectaris basins, and the anomalously thin lithosphere probably was penetrated during the impact which formed the Crisium basin. Such penetration is inferred from the morphologic evidence for internal modification seen in the highlands surrounding the basin.

Because of the advanced state of degradation of the basin, the original crater of Crisium cannot be identified by the morphologic evidence of pre-basin topography. Instead, the compositional data from orbital measurements and Luna 20 petrology suggest that excavation at Crisium was no deeper than upper and middle crustal levels, as was inferred for the Nectaris basin (see Chapter 4). A lower bound for the diameter of the transient cavity is probably about 380 km, the diameter of the inner mare ridge ring. If the transient cavity had been much smaller than this, the topography of the inner basin would be more prominent within Mare Crisium and terra islands would be evident within the mare fill of the basin. The maximum size of the transient cavity is difficult to constrain but is probably less than about 550 km in diameter. Such a value corresponds with the prominent massif ring (Figure 5.1) that bounds the mare fill of the basin. If the impact cavity had been much larger than this, mafic material from the lower crust and mantle of the Moon would have been excavated in contradistinction to the geochemical evidence. Moreover, compositional data from orbit and the Luna 20 site suggest basin excavation from middle to upper levels of the crust, about 40 to 45 km depth at most. Such depths of excavation imply a transient cavity diameter of about 450 km; there is currently no ring of this dimension evident in the Crisium basin.

The lack of an identifiable structure representing the original crater is explained by the basin modification model described by Schultz (1979) and Schultz *et al.* (1981). For basins that penetrate the rigid, outer layer of the Moon (lithosphere), sub-lithospheric flow immediately following the impact may partly or totally destroy any morphological remnant of the original crater of the basin (Schultz, 1979). According to this model, basin rings are megaslumps, but their position and morphology have been changed by internal modification. This endogenic model for basin rings was suggested by Hartmann and Wood (1971) for all basins, but was subsequently discounted on the basis of morphologic evidence from the Orientale example (Head, 1974a; Howard *et al.*, 1974). In the Crisium region, however, platform massifs and buried mare deposits (Figures 5.3, 5.4) suggest that this basin has undergone considerable structural modification accompanied by copious volcanic activity immediately following the basin impact. An initially small transient cavity for the Crisium basin (less than 500 km in diameter) may have been obliterated by the long-term internal modification, evidenced by platform massifs and mare basalt flooding within and around the Crisium ring troughs.

Platform massifs are abundant around the Crisium basin and I believe that they provide evidence for a distinct style of basin modification. These features result from penetration of the lithosphere by a basin-forming impact (Figure 5.12). The impact induces radial and concentric fractures in the rigid lithosphere (Figure 5.12a); such fracturing has been modeled analytically (Melosh, 1976) and documented on the Moon by observation (Mason *et al.*, 1976). After the impact, the lunar asthenosphere flows inward to compensate for the mass of crustal material excavated from the basin cavity. Because the viscosity of the asthenosphere is variable laterally on small scales, the regional inward flow produces platforms by selective removal of underlying support for the lithosphere (Figure 5.12b); failure occurs along zones of weakness induced earlier. Such a mechanism for the production of platform massifs would explain their absence around young basins, such as Orientale; these basins formed after the lunar lithosphere had grown to such a thickness that the transient cavity formed entirely within the rigid lithosphere (Solomon and Head, 1980).

The proposed model for the long-term, internal modification of the Crisium basin also may explain the discontinuous nature of the outer basin rings. In places, outer rings are prominent and display scarp-like morphology, but may be totally absent in many sectors (Wilhelms and El-Baz, 1977; Wilhelms, 1987). Moreover,

Figure 5.12 Schematic drawing showing the postulated sequence of events for the development of platform massifs. Sub-lithospheric flow toward basin center causes lithosphere to fracture along radial and concentric zones of weakness, previously established at time of basin impact.

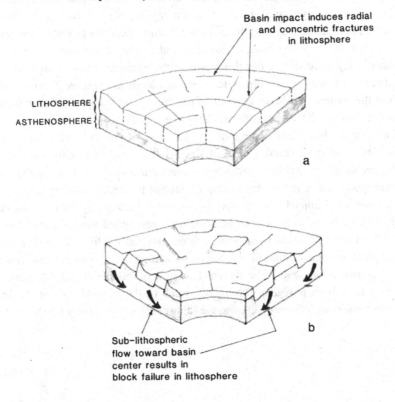

the 1080-km-diameter ring is not associated with a change in ejecta facies as is the main topographic rim in other basins (cf. Crisium rings and the Cordillera scarp of Orientale; Chapter 3). In my view, these large outer rings of the Crisium basin are structural features produced during long-term adjustment of the lithosphere to the smaller transient cavity of the Crisium basin. Support for this hypothesis comes from the observation that buried mare basalt deposits are found in the vicinity of these ring segments, as evidenced by the presence of dark-halo craters (Schultz and Spudis, 1979). I believe that a comparable process is responsible for the development of the outer rings of the Nectaris, Serenitatis, and Imbrium basins.

The ejecta from the Crisium basin consist of upper crustal anorthosites and anorthositic gabbros and low-K Fra Mauro basalts derived from middle to lower levels of the crust. The anorthositic rocks dominate the ejecta while LKFM constitutes about 20 percent or less. There is no evidence for sub-crustal lithologies in either the orbital data or the Luna 20 samples. Pure KREEP basalt is absent, as it is for most of the eastern hemisphere of the Moon (Metzger *et al.*, 1977), and the KREEP component in the Crisium region is probably contained in the melt ejecta of low-K Fra Mauro basalt composition. Mare basalt may occur within the ejecta, but post-basin volcanic flooding of the interplatform regions makes this conclusion uncertain.

Impacts occurring after the Crisium basin have had minor influence on regional geological patterns. Both the Imbrium and Serenitatis impacts probably contributed ejecta and secondary craters to the Crisium region and the smooth Cayley-like plains that mantle the ring troughs and interplatform deposits probably are derived in part from these distant basins. The hilly and furrowed material in the Crisium highlands may be related to distal deposits of the Serenitatis basin that have partly mantled the rim of post-Crisium craters such as Cleomedes (Figure 5.3a). However, most of this material probably consists of primary ejecta from the Crisium basin, as it does at the Luna 20 landing site near the crater Apollonius.

The Crisium basin has undergone a geological evolution unlike that of other basins described in this book. Evidence for long-term internal modification of lunar landforms has previously been restricted to small craters modified during the epochs of mare volcanism. The possible effects of internal processes in the modification of large, basin-sized impact cavities may be profound, particularly for older basins on the Moon. Such basins may not have originally resembled the Orientale, the supposed "archetype" of lunar multi-ring basins. Internal modification processes are also evident in the evolution of the Nectaris basin and for portions of the Imbrium and Serenitatis basins, as will be shown. This style of modification of basin topography may have been prevalent throughout much of the early history of the Moon, when thermal gradients were high and the lithosphere still relatively thin.

6

A transitional basin: Serenitatis

The Serenitatis basin is on the near side of the Moon, east of Mare Imbrium and north of Mare Tranquillitatis (Figure 1.1). The basin is almost completely flooded by mare basalts (Figure 6.1) and displays a mascon gravity anomaly (Sjogren *et al.*, 1974). The Serenitatis basin was recognized as multi-ring in the studies of Hartmann and Kuiper (1962), Baldwin (1963), Hartmann and Wood (1971) and during systematic geological mapping of the Moon (Wilhelms and McCauley, 1971). Because of the large amount of mare flooding and generally degraded appearance of the basin, Serenitatis was once considered to be one of the oldest basins on the Moon (Hartmann and Wood, 1971; Wilhelms and McCauley, 1971). This view has changed, primarily because of ages obtained for some Apollo 17 samples considered to represent impact melt of the Serenitatis basin (James *et al.*, 1978; Wilhelms, 1987). I will describe the regional geology of the Serenitatis basin and some aspects of Apollo 17 site geology that relate to problems in the interpretation of its formation and subsequent evolution.

6.1 Regional geological setting and basin definition

The Serenitatis basin is close to the Imbrium basin and the effects of Imbrium on the morphologic evolution of Serenitatis have been significant. Most interpretations of basin geology rely on the well exposed highlands to the east of Mare Serenitatis (Figure 6.1). Thus, the morphological data available for interpreting the geology of the Serenitatis basin are limited compared with those for some of the other basins described in this book. The basin impact occurred within a highlands crust of about 60 km thickness (Bills and Ferrari, 1976). The basin is only slightly younger than the Imbrium impact (Wilhelms, 1987) and mare volcanism was active at the time of Serenitatis basin formation (Ryder and Taylor, 1976; Ryder and Spudis, 1980). Thus, as at Crisium and Nectaris, the large amounts of heat inferred to be in the crust provided a mechanism to endogenically modify the original morphology of the basin.

Several older basins are found near Serenitatis (Figures 2.8, 6.2). To the south, the Tranquillitatis basin is intersected by outer rings of the Serenitatis basin

(Wilhelms and McCauley, 1971). This non-tangential intersection may be responsible for the topographic low that enabled the basalts of Mare Tranquillitatis to partly
embay the southern edge of the Serenitatis basin. In addition, the eastern rim of the
Serenitatis basin may owe its topographic prominence to the coincident alignment
of this segment of its ring system with the large, outer rings of the Imbrium basin
(Chapter 7).

One of the most unusual features of Serenitatis is its double-basin configuration.
Scott (1974) found that a gravity low or saddle separates the northern and southern
portions of the Serenitatis mascon. Photogeological interpretation of peculiarities of
the basin ring system (Scott, 1972b) and the presence of this gravity saddle suggest
that the Serenitatis "basin" is actually two basins, designated North and South
Serenitatis (Scott, 1974). The northern basin appears to be significantly smaller
(about 400 km in diameter) than the younger, southern basin (main rim diameter
920 km; Table 2.1). I use the term "Serenitatis" in this book to refer to the major
(and multi-ringed), southern basin of the Serenitatis pair.

Figure 6.1   Basin-centered view of the Serenitatis basin.   "Vitruvius Front"
(VF) represents main rim of basin Rectified photograph from the Lunar and
Planetary Laboratory, University of Arizona.

The double-basin configuration of Serenitatis has been both questioned (Wood and Head, 1976) and mapped somewhat differently (Maxwell *et al.*, 1975). As both photogeological and geophysical data suggest the existence of North Serenitatis, the double-basin structure seems well supported by the data. Maxwell *et al.* (1975) mapped the distribution of mare ridges within Mare Serenitatis and inferred the northern basin as being centered significantly to the northeast (center at about 36° N, 26° E) of the postulated position mapped by Scott (1974) at about 32° N, 16° E (Figure 6.2). This rendering makes the northern basin non-coincident with both the gravity data and photogeologic interpretation of the northern Serenitatis region and for these reasons, I prefer the reconstruction of Scott (1974). The implications of the double-basin configuration for Serenitatis geology are primarily for pre-existing topography and the contributions of crustal material from the older, northern basin to the impact site of the southern basin.

Figure 6.2 Map showing the double-basin configuration of Serenitatis, including the pre-existing North Serenitatis basin (Scott, 1974). Intersecting rings of the older Tranquillitatis basin to the south also shown.

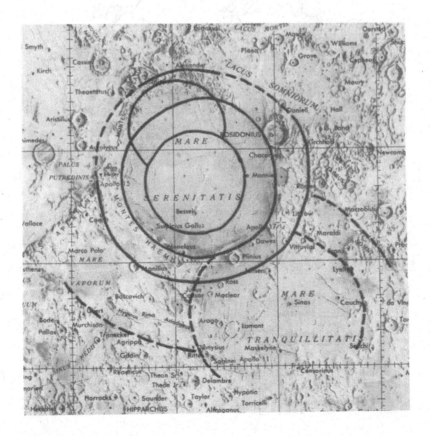

6.2 Serenitatis morphology and geological units

As the interior of the Serenitatis basin is extensively flooded by mare basalts, there are no exposures of geological units analogous to those contained within the Orientale, Nectaris, or Crisium basins. Presumably, a basin impact melt sheet underlies the basalts of Mare Serenitatis, and the terra surrounding the inner basin

Figure 6.3 Geological sketch map of the Serenitatis basin, after Wilhelms and McCauley (1971) and new mapping by author. Destructive effects of younger Imbrium basin have nearly obliterated the morphology of the deposits (but not ring) on western and northern sides of basin; plateaux in this area are largely related to Imbrium, but some may be Serenitatis remnants. Base is Lambert equal-area projection of lunar near side (Rükl, 1972).

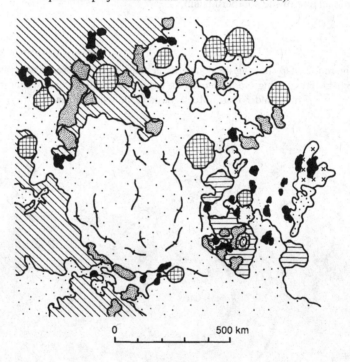

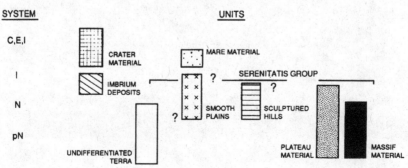

mare may be discontinuously mantled by the distal margins of such a melt sheet; this is postulated to have happened at the Apollo 17 landing site (see below). Most of the discussion in this section will relate to the well exposed terra east of the central basin mare whose morphology is probably influenced mostly by Serenitatis (Figures 6.3, 6.4), although the degradational effects of the younger Imbrium basin may have obliterated or modified primary Serenitatis texture in many areas.

The highlands east of the central Serenitatis basin are composed of two main terrain types (Figure 6.4): rugged massifs that are crudely aligned along basin-concen-

Figure 6.4 Oblique view of eastern rim of Serenitatis basin. Sculptured Hills (SH) are gradational with Imbrium basin lineated terrain (upper arrow). Smooth plains (P) appear to post-date Serenitatis basin. Apollo 17 landing site (A) near bottom. Note deposit of Sculptured Hills material on southern slope of South Massif (lower arrow); this could be the source of aphanitic breccias collected at South Massif stations at Apollo 17. North at top, field of view about 200 km at bottom. AS17-M-0938.

tric circles and a knobby terrain (Figures 6.3, 6.4) that has been called informally the Sculptured Hills (Scott and Carr, 1972; Wolfe *et al.*, 1982). The rugged massifs resemble the equant, isolated massifs that make up the Outer Rook ring of the Orientale basin (Figure 3.4) and they appear to form two distinct rings outside the flooded interior of the basin (Figure 6.5). The intermassif regions are undulating terra that has both rough and smooth facies. Some of the smooth regions resemble Cayley plains deposits, similar to those around the Crisium basin (interplatform plains; Figure 5.4). The hummocky intermassif deposits are gradational with the Sculptured Hills unit. All of these terra units are found near the Apollo 17 landing site (Figure 6.4).

A subject of considerable controversy is the source of material for the Sculptured Hills. Most investigators interpret the Sculptured Hills as an analog to the knobby, Montes Rook Formation of the Orientale basin (Wolfe and Reed, 1976; Head, 1979; Wolfe *et al.*, 1982). In this view, the Sculptured Hills are primary ejecta from

Figure 6.5 Ring interpretation map of Serenitatis basin. Main rim of basin (long dashes) is not the ring bounding the mare, but is expressed by Vitruvius Front, as shown from regional topography; main rim 920 km diameter. Other rings shown (see Table 2.1 for diameters); outer basin rings extend into rim of Imbrium basin and are partly responsible for topographic prominence of the Apennine Bench (Chapter 7). Base is Lambert equal-area projection centered on 26° N, 18° E (basin center).

the Serenitatis basin, deposited between the main topographic rim and the interme-
diate rings of the basin. In such a view, there was no subsequent burial of the
Sculptured Hills by debris of the impact which formed the Imbrium basin. The
Sculptured Hills unit appears to be gradational with lineated terrain just north of the
Apollo 17 site (Figure 6.4). This lineated terrain is radial to both the Imbrium and
Serenitatis basins (Scott and Carr, 1972), but occurs inside the basin rim of
Serenitatis, as interpreted by Wolfe *et al.* (1982).

A key relation that is not consistent with a Serenitatis origin for the Sculptured
Hills unit is its relationship to craters that post-date the Serenitatis basin, such as Le

Figure 6.6 Southwestern rim of Serenitatis basin. Ejecta from Imbrium basin
(lineated terrain) obscure deposits of Serenitatis, but mare-bounding rim is still
evident. Platform massifs (M) buried by Imbrium deposits. North at top,
framelet width 12 km. LO IV-90 H2.

Monnier and Littrow (Figure 6.4). These craters are superposed on the rings of Serenitatis, yet the Sculptured Hills material overlies the rims of these craters; such a stratigraphic relation is similar to that of the hilly materials that overlie the rim of Cleomedes at Crisium basin (Chapter 5). Spudis and Ryder (1981) interpreted these relations to mean the Sculptured Hills unit cannot be primary ejecta from the Serenitatis basin. The Sculptured Hills unit occurs throughout the eastern rim of the Serenitatis basin and appears to be gradational with the Alpes Formation of the Imbrium basin to the north of Mare Serenitatis. The Sculptured Hills unit may be polygenetic, in part composed of Serenitatis ejecta where it grades into hummocky, intermassif deposits (near the Apollo 17 site; Figure 6.4) and partly Imbrium ejecta where it is gradational with radially lineated terrain found inside the main rim of Serenitatis (Figure 6.4).

The western rim of Serenitatis is largely obscured by the pervasive effects of the

Figure 6.7 Platform massif on north rim of Serenitatis basin; this massif occurs on the Vitruvius ring of the basin and is partly mantled with knobby Alpes Fm. material from the Imbrium basin. North at top, framelet width 12 km. LO IV-91 H3.

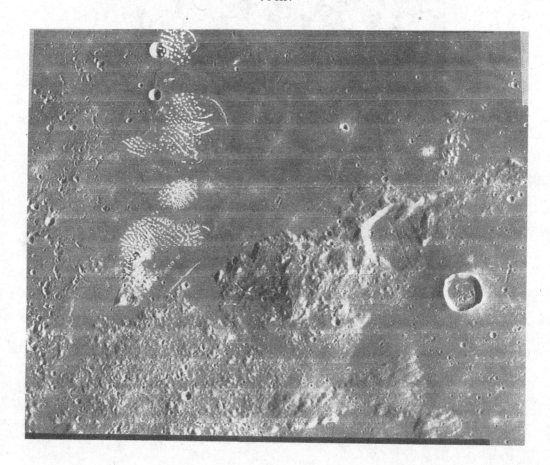

Imbrium impact. One characteristic landform of this region that was probably created by the Serenitatis impact is the platform massif. Found around the main basin rim, in the Montes Haemus (Figure 6.6), near Vitruvius, south of the Apollo 17 landing site, and north of Mare Serenitatis (Figure 6.7), these massifs are similar in appearance to those around the Crisium basin, but less abundant. I interpret platform massifs as structural features that form in response to relaxation of the lunar crust after basin excavation (sec. 5.6).

Serenitatis displays no recognizable textured ejecta deposits outside of its main rim. The lack of ejecta in part reflects the extensive deposition of Imbrium ejecta and secondary cratering (Figure 6.3), but such degradational effects are probably not the sole explanation for this puzzling absence. The distal deposits of the Serenitatis basin are exposed primarily northeast of the basin where secondary craters interpreted to be from Serenitatis appear to overlap deposits of both the Humboldtianum and Crisium basins (Wilhelms, 1976a). If this relation is correct, Serenitatis must post-date the Crisium basin, but pre-date the Imbrium basin (Wilhelms, 1981, 1987).

There is no requirement that Serenitatis be an extremely old basin based solely on the heavily cratered appearance of its eastern rim. The eastern rim of Serenitatis appears to be scarred by abundant, overlapping craters of large (50 km) size (Figure 6.1). This heavily cratered texture probably is responsible for interpretations that Serenitatis is one of the oldest basins on the Moon (Hartmann and Wood, 1971; Wilhelms and McCauley, 1971). The distribution of geological units in this region (Figure 6.3) suggests an alternative explanation for this appearance. The gradational relations seen between the Sculptured Hills and the radially textured deposits in the southeast and Alpes Formation of the Imbrium basin in the northern sectors of the basin suggest that at least some of this material is ejecta from the Imbrium basin. Wilhelms (1980a) suggests that the heavily cratered eastern rim of the Serenitatis basin is a product of extensive secondary cratering from the Imbrium basin; these craters are overlain by the Sculptured Hills and related highland units. I believe that some of the large craters in the eastern Serenitatis rim region (e.g., Littrow; Figure 6.4) may be Imbrium secondaries that were formed contemporaneously with the deposition of distal ejecta from Imbrium. Thus, some of the highland units on the eastern rim may both underlie and overlie these large craters, producing ambiguous stratigraphy; such relations are seen in the secondary craters of the Orientale basin (Moore *et al.*, 1974; Wilhelms, 1976a; Schultz and Mendenhall, 1979).

6.3 Serenitatis basin rings and structure

Serenitatis displays at least four distinct rings (Table 2.1; Figure 6.5), with the suggestion of a faint outer ring in the highlands south of Mare Vaporum (Wilhelms and McCauley, 1971; Head, 1974c, 1979). The inner (Linné) ring is not expressed by basin massifs, but its presence is implied by the circular arrangement of mare ridges within the central mare basin. This ring is about 410 km in diameter and is probably analogous to the mare ridge ring seen at both the Nectaris and Crisium basins. The main highland ring that borders the mare is composed of rugged massifs and basin-facing scarps and is approximately 620 km in diameter. This ring passes through the Montes Taurus east of the basin and is interpreted as the rim of the transient cavity of the Serenitatis basin by Head (1979). The Taurus ring passes just west of the Apollo 17 landing site and intersects the northern Apennines of the Imbrium basin; this relation has important implications for the geology of the Apollo 15 landing site (sec. 7.6). The main Serenitatis basin rim is defined by highland scarps and isolated massifs as well as the "Vitruvius Front", a well-defined platform massif southwest of the Apollo 17 site (Figure 6.1). The Vitruvius ring is about 920 km in diameter and corresponds to the Cordillera scarp of Orientale according to Head (1979) and Wolfe *et al.* (1982). That no textured ejecta are evident in association with this basin ring may result from the extensive obliteration of Serenitatis ejecta by the effects of the impact forming the Imbrium basin.

Outside the main basin rim, two additional rings are evident (Figure 6.5). These rings are similar in morphology to the outer rings of the Orientale (Chapter 3) and Imbrium (Chapter 7) basins; they have diameters of 1300 and 1880 km, respectively (Pike and Spudis, 1987). Parts of these rings can actually be traced inside the Apennine ring of the Imbrium basin (Figures 2.7, 6.2), a relation that has important implications for reconstructing the size of the transient crater of the Imbrium basin, as discussed in Chapter 7.

The rings of the Serenitatis basin display a crudely polygonal outline that reflects interaction with topography both older and younger than the basin. The outer Vitruvius ring is missing where the Serenitatis basin intersects the Tranquillitatis basin in the south and is truncated by the Apennines of the Imbrium basin to the west. In addition, the Vitruvius ring becomes highly irregular northeast of the central basin and a tangential structure is evident near the crater Atlas. These relations are similar to the irregularities of the Cordillera ring of Orientale (Chapter 3) and are consistent with the interpretation that the Vitruvius ring is the main topographic rim of the Serenitatis basin.

As at the Crisium basin, the intermediate (Taurus) ring of Serenitatis displays platform massifs (Figure 6.6) as does the main (Vitruvius) ring (Figure 6.7). These platform massifs are probably related to structural modification of pristine basin topography, although the effects of this style of modification were not as pervasive as they are at Crisium. Thus, it appears that the Serenitatis basin is analogous in size and morphology to the Nectaris basin, the principal differences being the more extensive degree of mare flooding and destructive effects of the Imbrium impact on Serenitatis ejecta and morphology.

6.4 Orbital geochemical data for Serenitatis basin deposits

Information on the regional composition of Serenitatis deposits is available only for the southeastern corner of the basin rim. This meager coverage is a result of both

Table 6.1 *Chemical composition and mixing model results for the Taurus–Littrow highlands, Serenitatis basin*

Region	Al wt.%	Mg wt.%	Fe wt.%	Ti wt.%	Th ppm	FAN %	NOR %	TROC %	KREEP %	11MB %
Taurus–Littrow	9.2	6.0	6.5	1.7	1.4	1.2	67.7	2.2	2.4	26.5
Taurus–Littrow (normalized)	–	–	–	–	–	1.6	92.1	3.0	3.3	–

Note: Region Taurus–Littrow defined in Spudis and Hawke (1981), roughly consists of area shown in Figure 6.4. Mixing model results from Spudis and Hawke (1981). Normalized values assume zero mare basalt component (see text).

the extensive mare basalt flooding that masks the inner basin deposits and the effects of Imbrium ejecta in the western rim area that cover most, if not all, of the Serenitatis deposits in that area. The Taurus–Littrow highlands are located between the Taurus and Vitruvius rings of Serenitatis and are partly embayed by mare basalts and discontinuously mantled by pyroclastic deposits associated with mare flooding (Heiken *et al.*, 1974). The orbital X-ray data are of sufficient resolution to distinguish these deposits (Schonfeld, 1980) and the Serenitatis highlands units may be characterized chemically without the model-dependent subtraction of a mare basalt component. However, mixing model studies of orbital chemistry require gamma-ray information, which is of low resolution (Adler and Trombka, 1977). Therefore, the calculations presented here used all available chemical data, from which the mare component was then subtracted and the values normalized to 100 percent. This assumes that mare basalt is absent as a component of Serenitatis basin ejecta. Mare basalt is present in quantities of less than one percent in pure highlands soils from the Apollo 17 site (Ryder, 1981b) and this recalculation is thought to have no significant effects on our estimate of the petrological composition of the highlands (Spudis and Hawke, 1981).

The regional composition of the Taurus–Littrow highlands is distinctly different from the compositions seen around the Crisium and Nectaris basins (Table 6.1). Examination of individual pixels of X-ray data (Spudis and Davis, 1983) shows that unmantled highlands in this region display consistently lower Al/Si and higher Mg/Al values than do the terra surrounding the Nectaris basin. Regional gamma-ray values for Fe and Ti are considerably higher, although this is caused mostly by the embayment of these highlands by high-Ti mare basalts (Scott and Carr, 1972). The Taurus–Littrow highlands are geochemically distinct from any other highlands region associated with basins on the Moon (Spudis and Hawke, 1981).

Results of mixing model calculations for this region are presented in Table 6.1. These results are in striking contrast to the composition of ejecta from the Nectaris basin (cf. Tables 4.1, 6.1). Norite is the dominant rock type, constituting over 90 percent of the basin deposits overflown by the Apollo orbital instruments. Anorthosite appears to be a minor component (abundance less than 2 percent), in accord with the results of studies of samples from the Apollo 17 landing site which indicate that ferroan anorthosite is virtually absent from the Taurus–Littrow highlands (Ryder, 1981b). Both troctolite and KREEP are present (about 3 percent each); the KREEP component is probably Serenitatis basin impact melt entrained within the ejecta deposits, represented at the Apollo 17 site by the impact-melt "sheet" (Spudis and Hawke, 1981). The mare component was subtracted in the normalized results, but it may be present in the basin deposits, probably on the order of about one percent.

The low levels of KREEP in the mixing model results may be caused by a number of factors. Melt rocks rich in KREEP are abundant in the Apollo 17 collection (Simonds, 1975; Winzer *et al.*, 1977), but sample collection at the site concentrated on the boulders at the base of the massifs (AFGIT, 1973); thus, these rocks are probably over-represented in the sample collection. Moreover, the regional Th value used in the mixing model calculation, which strongly controls the total KREEP abundance, was taken from the compilation of Metzger *et al.* (1977) and these data include Th values for a number of geological units because of the low resolution of the gamma-ray instrument. Therefore, the total KREEP abundance of the Serenitatis basin deposits in this region may be slightly higher than is suggested by the results of the mixing model calculations (Table 6.1).

These results have important implications for the geology of the Serenitatis basin as well as the geology of the lunar highlands as a whole. The impact forming the Serenitatis basin occurred into a target largely composed of Mg-suite rocks; apparently, ferroan anorthosite was absent. This suggests either that an anorthosite layer was never formed in this region of the Moon (Ryder, 1981b) or that such a layer was removed by some process prior to the impact which created the Serenitatis basin. One possibility is that the crust in this region was intruded by an Mg-suite batholith early in lunar history, after solidification of the global magma ocean (James, 1980). Alternatively, the anorthosite-rich layer of the crust may have been removed from the Serenitatis target by one or several of the many older basins. Whatever the explanation for this distinctive composition, the orbital chemical data suggest that the impact targets were different for each lunar basin and that "typical highlands crust" may not be a realistic concept; each region of the lunar highlands has undergone a distinct petrological and geochemical evolution. Such distinct histories produce unique geochemical provinces within the lunar highlands (Spudis and Hawke, 1981).

Mare volcanism was probably active in the region, prior to the basin impact. This activity, and its associated high thermal gradient, may provide a mechanism for the internal modification of basin morphology, as suggested by the platform massifs of the basin (Figures 6.6, 6.7). The lunar surface may have been partly covered by basalt flows at the time of the basin impact, resulting in the inclusion of basalt fragments within the clastic ejecta of the basin.

Table 6.2 *Composition of melt rocks from Taurus-Littrow highlands,*
Apollo 17 landing site

Wt. %	1[a]	2[b]	3[c]	4[d]	5[e]	6[f]
Ti	0.88	0.87	0.54	0.61	0.84	0.46
Al	9.5	9.1	11.2	9.6	8.53	10.3
Fe	7.0	7.0	6.57	7.57	7.2	6.6
Mg	7.4	7.6	6.4	7.0	10.3	5.22
ppm						
Th	6.0	—	5.0	4.0	—	—
Sc	16	17	19	21.3	14	—
Sm	15	15	14.5	12	11	—

[a]Station 6 boulder, poikilitic melt rocks (LSPET, 1973)
[b]Station 7 boulder, poikilitic melt rocks (Winzer *et al.*, 1977)
[c]Boulder 1, station 2 aphanitic melt rocks (Blanchard *et al.*, 1975)
[d]Loose sample 73255 aphanitic melt rock (James *et al.*, 1978)
[e]Loose sample 76055 poikilitic melt rock (Albee *et al.*, 1973)
[f]Rake sample 72558, KREEP-rich melt rock (Warner *et al.*, 1977)

6.5 Apollo 17 site geology – the Serenitatis basin "melt sheet"

One of the primary objectives of the Apollo 17 mission to Taurus–Littrow was to sample the highland materials that make up the Serenitatis basin deposits in this region of the Moon. Sample collection at the landing site concentrated on boulders that appear to have originated on the upper slopes of the massifs at the site (Schmitt, 1973). These highland samples are the best direct source of data on the ejecta produced by the Serenitatis basin impact.

The most abundant rock type returned from the Apollo 17 highlands are impact melts having a composition of low-K Fra Mauro basalt (Simonds *et al.*, 1974; Simonds, 1975; Dymek *et al.*, 1976; Winzer *et al.*, 1977; Spudis and Ryder, 1981). While the melt rocks display a variety of textures and grain sizes, including sub-ophitic (basaltic) texture (Figure 6.8); the most common texture is *poikilitic* (in which pyroxene grains completely enclose plagioclase grains) and these rocks are often referred to as the "poikilitic melt rocks" (e.g., Spudis and Ryder, 1981).

The poikilitic rocks are abundant at the site and make up the entire Station 6 and 7 boulders, boulders 2 and 3 at the South Massif Station 2, and discrete samples found at all highland stations (Wolfe *et al.*, 1982). They contain abundant clasts from the lower levels of the lunar crust (Simonds, 1975; Dymek *et al.*, 1976), including norite, troctolite, and dunite; rare fragments of mare basalt indicate some upper crustal material was also included. The LKFM composition of the Apollo 17 melts is grossly similar to poikilitic melt rocks found at the Apollo 16 site, although the Apollo 17 samples appear to be slightly more mafic (Table 6.2). The petrologic,

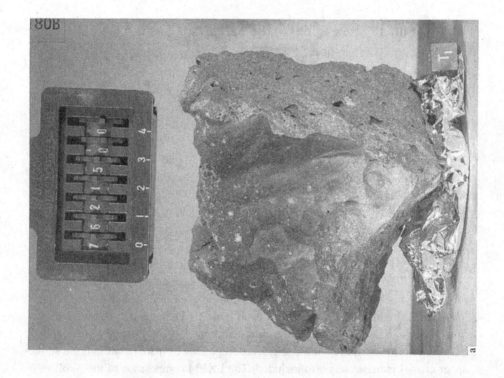

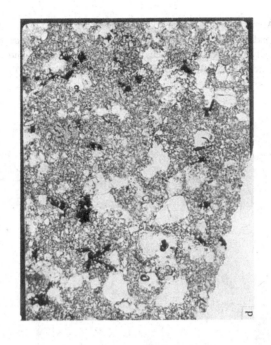

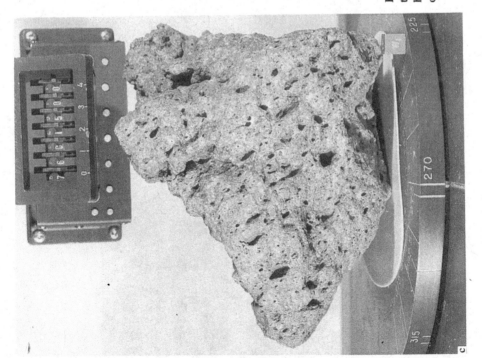

Figure 6.8 Poikilitic impact melt rocks 76215(a, b) and 76015(c, d) from the Apollo 17 landing site. These rocks show remarkable chemical homogeneity and petrographic diversity and are interpreted as fragments of the "melt sheet" of the Serenitatis basin. See Spudis and Ryder (1981).

Figure 6.9 Aphanitic impact melt rocks 72215(a, b) and 73255(c, d) from the
Apollo 17 landing site. In contrast to the poikilitic rocks (Figure 6.8), these
samples display chemical differences from the "melt sheet" and from each
other. Aphanites also show accretionary structures, suggesting they are melt

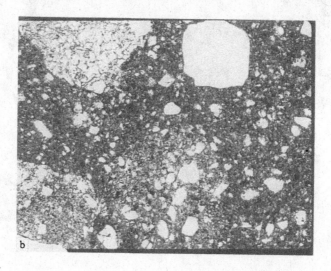

ejecta, analogous to terrestrial suevites (subsec. 2.1.2). Sample 73255 has a vesicular rind of melt, as evidence by vesicles on opposing side of the rock (V); these relations suggest it is a melt bomb. These impact melts may or may not be related to the Serenitatis basin impact. See Spudis and Ryder (1981).

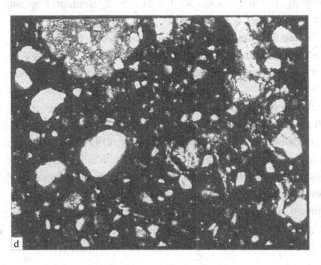

geochemical, and geological characteristics of the Apollo 17 poikilitic rocks were held to be analogous to the impact melt sheet seen at the terrestrial Manicouagan crater (Simonds *et al.*, 1976; Phinney *et al.*, 1978) and interpreted to be fragments of the impact melt sheet of the Serenitatis basin (Winzer *et al.*, 1977). The model of Wolfe *et al.* (1982) envisions a sequence of melt lenses that are entrained within a more clastic ejecta deposit that makes up the Taurus–Littrow massifs.

Not all of the poikilitic melt rocks have the same composition. The discrete sample 76055 (Warner *et al.*, 1973b; Albee *et al.*, 1973) has a significantly higher Mg content (Table 6.2) which may indicate that it was formed in a separate impact event (Spudis and Ryder, 1981). Several melt breccias among the Station 2 rake samples are significantly more KREEP-rich than the LKFM composition of the main melt sheet (Table 6.2; Warner *et al.*, 1977). These rocks may represent large impacts other than the one that formed Serenitatis; one possibility is that they are melt rocks from earlier basins, emplaced at the Apollo 17 site as clastic ejecta from the Serenitatis basin.

Other impact melts collected at the Apollo 17 site are the aphanitic melt breccias (Ryder *et al.*, 1975; James *et al.*, 1978; Spudis and Ryder, 1981). These rocks are extremely fine-grained (aphanitic; Figure 6.9) and include clasts of mare basalt, felsite, and norite, as well as abundant mineral debris (Ryder *et al.*, 1975; James *et al.*, 1978). The aphanitic breccias not only display compositions distinctly different from the poikilitic melt rocks (Table 6.2), but they also display considerable compositional variation from each other (Spudis and Ryder, 1981). The clast assemblage is also distinct from the poikilitic rocks, being primarily shallow-level crustal rocks, such as mare basalts, a KREEPy basalt, and felsite; they contain few troctolites and dunites that are abundant in the poikilitic breccias (Simonds *et al.*, 1974). The range in assembly ages for the Apollo 17 breccias suggests that while the poikilitic rocks all appear to have been formed at about the same time (approximately 3.87 Ga ago; Winzer *et al.*, 1977), the aphanitic melts display a range of crystallization ages from 3.93 Ga for the Station 2, Boulder 1 aphanites (Leich *et al.*, 1975) to the younger 3.86 Ga ages for the Station 3 sample 73255 (Eichhorn *et al.*, 1979). These differences led Spudis and Ryder (1981) to suggest that the poikilitic and aphanitic impact melts from the Apollo 17 site were formed in a variety of impact events and only the poikilitic melts are from the melt sheet of the Serenitatis basin.

The poikilitic rocks probably represent the melt phase of Serenitatis basin ejecta at the Apollo 17 landing site. This melt phase may be part of a melt "sheet" as postulated by Simonds *et al.* (1976) and Winzer *et al.* (1977), or ejected melt contained as pods within a clastic sequence as described by Wolfe *et al.* (1982). I prefer the interpretation that the poikilitic rocks represent the distal margins of the melt sheet of the Serenitatis basin and that they discontinuously mantle the massifs in this region (Figure 6.10). This interpretation is more consistent with the field observations of Schmitt (1973) that the boulders sampled at Station 6 were derived from high on the North Massif, within a well-defined color unit.

The origin of the aphanitic melts is a problem. On the basis of strict analogy to the melt sheet of the terrestrial Manicouagan crater, differences between the aphanites and the poikilitic rocks in age, composition, and clast content suggest that the aphanites are from many impacts that both pre-date and post-date the Serenitatis basin; if so, their collection at the Apollo 17 site was fortuitous (Spudis and Ryder, 1981). The occurrence of other melt rocks of anomalous composition at the site (Warner *et al.*, 1977) supports this interpretation. On the other hand, differences between the aphanites and the poikilitic rocks are minor (James *et al.*, 1978); bulk composition of both is LKFM basalt, textures and petrographic structures can vary widely within melt sheets, and all impact melts on the Moon appear to have roughly the same age (around 3.85 Ga), thus giving rise to the idea of a cataclysm (e.g., Ryder, 1990). If the VHA impact melts found at Apollo 16 are melt ejecta from the Nectaris basin (and it is difficult to envision an alternative to such derivation; see sec. 4.4 and Spudis, 1984), then melt sheets on the Moon can be quite diverse in composition and the terrestrial analogy fails. I see no compelling basis to choose between these two possibilities and the origin of the Apollo 17 aphanites remains unknown.

Many fragments of plutonic rocks were returned from the Apollo 17 landing site both as individual samples and as clasts within the melt rocks. These samples are

Figure 6.10 Geologic cross section of Apollo 17 landing site. Serenitatis basin melt sheet probably mantles massifs near site, or may occur in part as discontinuous lenses within massifs. Aphanitic melts could be exotic to site. By the author in Heiken *et al.* (1991).

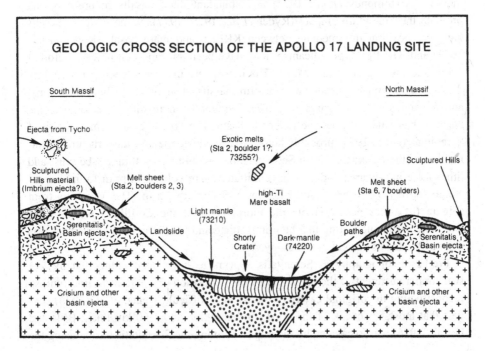

all derived from the Mg-suite (James, 1980) and no unequivocal ferroan anorthosite from Apollo 17 has been described to date. Norite is by far the most abundant rock type, found as clasts within the melt breccias (Ryder *et al.*, 1975) and as a large boulder at Station 8, which may or may not be representative of the Sculptured Hills (Jackson *et al.*, 1975). Hand samples of both troctolite (76535, 76536) and dunite (72415) were collected by Apollo 17 (Dymek *et al.*, 1975). This population is consistent with the interpretation derived from the mixing model studies that norite is the dominant rock type in the highlands surrounding the Serenitatis basin and that anorthosite is virtually absent. I interpret these plutonic fragments as clastic ejecta from the Serenitatis basin at the Apollo 17 landing site.

The results of the Apollo 17 mission permit detailed characterization of the ejecta produced by the impact that formed the Serenitatis basin. As at the Apollo 16 site, a two-component deposit was sampled consisting of a melt phase of LKFM composition and a clastic phase composed of Mg-suite plutonic rocks; anorthosite is conspicuously absent. Exotic material may be present at the site, possibly related to the arrival of ejecta from the Imbrium basin or from smaller, local craters (Spudis and Ryder, 1981).

6.6 The formation and evolution of the Serenitatis basin

The Serenitatis impact occurred into a highland region petrologically distinct from typical lunar crust. Mare volcanism probably was active at the time of the basin impact, as shown by the inclusion of rare clasts of mare basalt within the poikilitic melt rocks (Simonds *et al.*, 1974) and abundant mare basalts in breccias that pre-date the Serenitatis impact (Ryder *et al.*, 1975, 1977). Some of these basalts have composition intermediate between KREEP and mare basalt (Ryder *et al.*, 1977) and KREEPy-type volcanism may have been an important process before 4 Ga ago in this region of the Moon. The lithosphere may have been penetrated by the Serenitatis impact, producing platform massifs in segments of both the Taurus and Vitruvius rings (Figures 6.6, 6.7); Serenitatis more closely resembles the Nectaris basin than it does the Orientale basin. The double-basin configuration of Serenitatis (Scott, 1974) probably helped to accentuate the unusual structural modification of the basin; the North Serenitatis basin had already thinned the crust and lithosphere in this area and may have initiated mare volcanism prior to the impact which formed the larger, multi-ring South Serenitatis basin. The Vitruvius front (Figure 6.1) does not resemble the morphology of the Cordillera scarp of the Orientale basin, suggesting that both immediate and later structural adjustment was responsible for its current morphology.

Pre-basin topography at Serenitatis appears to have been obliterated by late structural adjustment of the basin and by the deposition of ejecta from Imbrium, so reconstructing the original cavity of the basin is difficult. Dence *et al.* (1976) suggest that most, if not all, of the crust (in this region about 60 km thick; Bills and Ferrari, 1976) was excavated by the impact forming the Serenitatis basin. Using the relationship between excavation depth and the diameter of the transient cavity described by Croft (1981a), such an excavation of the crust suggests a transient cavity 500 to 600 km diameter. Head (1979) proposes the intermediate Taurus ring as the transient cavity rim (620 km), while Wolfe *et al.* (1982) suggest a diameter of 750 km, corresponding to a zone between the Taurus and Vitruvius rings. If the excavation cavity were much larger than either of these values, mantle material should be present in the basin ejecta; apparently, it is not. If the cavity was significantly smaller, it is difficult to account for the large size of Serenitatis, a basin nearly the same size as Orientale (930 km diameter; Chapter 3). On the basis of such considerations, I estimate a diameter of 600 ± 100 km for the excavation cavity of the Serenitatis basin. The large uncertainty of this estimate is caused in part by the obscuration of Serenitatis geology by subsequent events.

The adoption of a cavity smaller than the one suggested by Wolfe *et al.* (1982) implies that the Apollo 17 site is *not* located on the rim of the basin transient cavity. Thus, the Apollo 17 "melt sheet" is probably a discontinuous mantle of melt deposits spilled from the margins of the main melt sheet, as seen by analogy at such smaller impact craters as Tycho (Shoemaker *et al.*, 1968). Alternatively, the Apollo 17 impact melts may be derived from ejected melt contained within the clastic sequence at the landing site as described by Wolfe *et al.* (1982). Because there is no direct evidence for melt ejection from the Apollo 17 poikilitic melt rocks, I prefer the former interpretation.

The Sculptured Hills unit is related to the Serenitatis basin, but in a complex manner. It is not directly analogous to the Montes Rook Formation of the Orientale basin as claimed by Head (1979) and Wolfe *et al.* (1982), because it overlies (and therefore, post-dates) craters that post-date the Serenitatis basin (Figure 6.4; Spudis and Ryder, 1981). Parts of the Sculptured Hills are gradational with hummocky deposits that are very likely of Serenitatis origin (Figure 6.4) and these highlands probably contain a large fraction of Serenitatis basin ejecta. The Sculptured Hills material that overlies the craters Littrow and Le Monnier must be derived mostly from the Imbrium basin; they resemble the morphology of the Alpes Formation of the Imbrium basin (see Chapter 7) and are gradational with both the Alpes Formation and radially lineated terrain north of the Apollo 17 site (Figure 6.4). In most areas of the eastern rim of the Serenitatis basin, the Sculptured Hills probably represent a complex mixture of ejecta from both the Serenitatis and Imbrium basins and smaller amounts of locally derived material.

The ejecta produced by the Serenitatis basin impact contrast strongly in composition with the ejecta of the Nectaris and Crisium basins. The Serenitatis deposits are norite, with lesser amounts of KREEP, mare basalt and other mafic rocks of the Mg-suite, derived from lower crustal levels. This dissimilarity indicates that the lunar highlands are petrologically heterogeneous and that basins formed within distinct geochemical provinces (Spudis and Hawke, 1981). Serenitatis basin ejecta are a two-component mixture of melt and clastic debris. The Apollo 17 poikilitic rocks represent the basin impact melt phase. These rocks are not abundant in the

Taurus–Littrow region according to the results of the mixing models of orbital chemical data and this probably reflects their distribution as a thin veneer discontinuously mantling the massifs near the Taurus ring. The aphanitic melt breccias collected at the Apollo 17 site could be related to a variety of impacts that both pre-date and post-date the Serenitatis basin; some may be derived from the Imbrium basin. Alternatively, the aphanites may be melt ejecta from the Serenitatis basin, possibly ejected from the crater early in the impact sequence (e.g., Wood, 1975).

We do not know the reason for the dominance of norite within Serenitatis ejecta. It may be that an anorthosite layer was never present in this region of the Moon and hence its absence within the ejecta of the basin (Ryder, 1981b). Another possibility is that the anorthositic crust was partly removed by the impacts of earlier basins, such as North Serenitatis. The crust was then intruded by magmas (James, 1980) that crystallized to form an Mg-suite pluton. Such a magma chamber could contain the cumulate dunite and troctolite sampled at the Apollo 17 landing site. Mare volcanism was active before and at the time of the Serenitatis impact and this material was probably included within the basin ejecta deposits.

The Serenitatis basin underwent structural adjustment after the initial modification stage of basin formation, as shown by the platform topography of portions of the Taurus and Vitruvius rings. The Imbrium impact destroyed most of the western rim of the Serenitatis basin and has probably affected large areas of its eastern rim by secondary cratering and ejecta deposition. Modification of Serenitatis topography was probably accentuated by the effects of the Imbrium impact and the proximity of Serenitatis to Imbrium may be responsible for its degraded appearance, an attribute that long has hampered understanding of the geology of the Serenitatis basin.

7

The largest basin: Imbrium

The Imbrium basin (Figure 7.1) is probably the most studied multi-ring basin on the Moon. Prominently located on the lunar near side, west of Mare Serenitatis and east of the large maria Oceanus Procellarum (Figure 1.1), the Imbrium basin first attracted the attention of G.K. Gilbert (1893) in his historic analysis of lunar craters. Gilbert recognized the extensive pattern of radial texture associated with Imbrium and postulated that Mare Imbrium had formed by the collision of a large meteorite with the Moon. The impact origin of the Imbrium basin was also recognized by Dietz (1946), Baldwin (1949; 1963), Urey (1952), and Hartmann and Kuiper (1962). The landmark paper of Shoemaker and Hackman (1962) proposed a global stratigraphic system for the Moon based on the deposition of ejecta from the Imbrium basin as a marker horizon. The Imbrium impact was considered such a key event in lunar geological history that two Apollo missions (Apollo 14 and 15) were sent to landing sites specifically chosen to address problems of Imbrium basin geology.

7.1 Regional geology and setting

Imbrium is one of the youngest major basins on the Moon, but extensively flooded by mare basalt (Figure 7.1). Even so, as one of the largest lunar basins (main topographic rim 1160 km in diameter), it has an ejecta blanket so extensive that almost all of the near side may be dated relatively with respect to the time of the Imbrium impact (Wilhelms, 1970). The site for the impact which formed the Imbrium basin appears to have been unusual in several important aspects. The crust in the Imbrium region is relatively thin, about 55 km or less in most places (Bills and Ferrari, 1976; Goins et al., 1979). Results of the orbital gamma-ray experiment (Adler and Trombka, 1977; Metzger et al., 1977) show that there are unusually high values of radioactivity in the Imbrium–Procellarum region. The regional composition suggests that ferroan anorthosite was not abundant in the target site for the Imbrium basin. Mare volcanism was active in pre-Imbrian time (Shoemaker, 1972; Ryder and Taylor, 1976; Schultz and Spudis, 1979, 1983) and flows of mare basalt probably covered at least part of the basin target. KREEP volcanism may have occurred in this region of the Moon (Schonfeld and Meyer, 1973; Meyer, 1977) and KREEP basalts may interfinger with mare basalt flows in this area.

There are many structures and topographic features older than Imbrium in the region (Wilhelms and McCauley, 1971; Spudis and Head, 1977; Head, 1977b). Whitaker (1981) proposed that the Imbrium impact occurred within an ancient and large basin, the Procellarum (or Gargantuan; Cadogan, 1974) basin. This basin, if

Figure 7.1 Basin-centered photograph of the Imbrium basin, showing prominent landmarks, the location of the Apollo 15 landing site, and pre-existing basins Serenitatis (Chapter 6) and Insularum (Table 2.2). North at top, width of scene about 1500 km. Photograph from Lunar and Planetary Laboratory, University of Arizona. From Spudis and Ryder (1985).

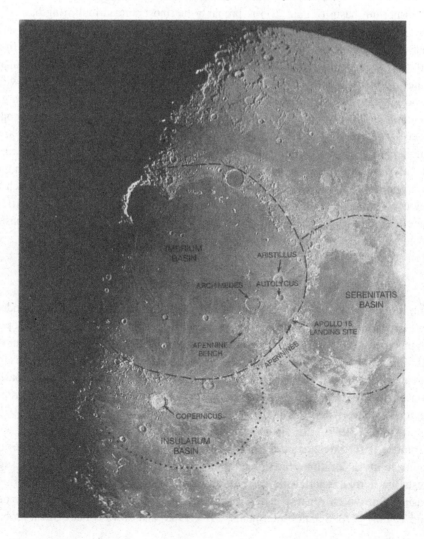

real, was over 2300 km in diameter and covered most of the near side of the Moon (Figure 7.2). Such a basin not only would have removed most of the pristine lunar crust in the Imbrium target area, but would possibly have served as a site for igneous activity. The presence or absence of the Procellarum basin has such important ramifications for the geology of the Imbrium basin, and indeed, for the entire Moon, that I critically examine the evidence for its existence below in the discussion of the ring system of the Imbrium basin (sec. 7.3).

Several other better-defined basins near Imbrium have affected the present morphology of the basin (Figure 7.2). The Insularum basin (Wilhelms and McCauley, 1971; Wilhelms, 1980a, 1987) intersects the outer Imbrium ring near Copernicus and may be responsible for the suppression of Imbrium topography in that region and the accentuation of inner basin topography in the Apennine Bench area (Head, 1977b; Spudis and Head, 1977). The Serenitatis basin, discussed in Chapter 6, also

Figure 7.2 Possible basins in the target region of the Imbrium basin that could have existed and affected Imbrium geology. Basin configurations: Serenitatis (Scott, 1974; Wolfe and Reed, 1976); Insularum (Wilhelms and McCauley, 1971); East Procellarum (De Hon, 1979); Procellarum (Whitaker, 1981). As discussed later in this chapter, Procellarum basin probably does not exist. Base by U.S. Geological Survey.

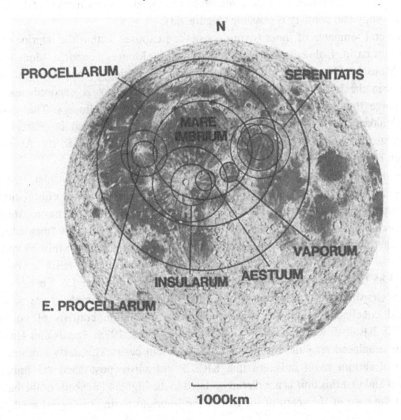

had important effects on the topography of the eastern rim of the Imbrium basin (Spudis and Head, 1977; Head, 1979). Imbrium and Serenitatis basin rings intersect at the Apollo 15 landing site, producing an uplifted section of pre-Imbrian rocks within the Apennine Front (sec. 7.6). Other nearby basins include the Vaporum and Aestuum basins, and a questionable basin in eastern Oceanus Procellarum (De Hon, 1979).

The lunar lithosphere appears to have been relatively rigid at the time of the Imbrium impact (Ferrari *et al.*, 1978; Solomon and Head, 1980). However, because of the thin crust in this area, abundant volcanic activity, and the large size of the Imbrium impact, sub-lithospheric creep after basin formation may have occurred in parts of the Imbrium area (Whitford-Stark, 1981a). Thus, I will consider the possibility of some post-impact, internal modification to the basin in the geological reconstruction of the history of the Imbrium basin (sec. 7.8).

7.2 Imbrium morphology and geological units

Because of the abundant mare flooding, the deposits of the Imbrium basin are not as completely exposed as those of the Orientale basin (Chapter 3). However, several parts of the basin interior have not been flooded, and the outer deposits are well preserved in many areas. Thus, a reasonably complete reconstruction of Imbrium morphology and geology is possible (Figure 7.3).

Several remnants of inner basin deposits are exposed within the interior of the Imbrium basin. Isolated massifs, as represented by Montes Teneriffe, Mons Piton, and Mons La Hire, occur along concentric circles outside the inner ring of mare ridges in the basin. In addition, a large highland remnant is exposed near the Apennine Bench (Figure 7.1), just south of the crater Archimedes. This rugged terra makes up part of the intermediate ring of the basin and is embayed by pre-mare light plains (Figure 7.1) that have been formally named the Apennine Bench Formation (Hackman, 1966).

Some workers (McCauley, 1977; Wilhelms *et al.*, 1977) consider that the Apennine Bench Formation is the Imbrium equivalent of the Maunder Formation of the Orientale basin, i.e., it is the impact melt sheet of the Imbrium basin. Another interpretation holds that the Apennine Bench Formation is a giant "megablock", contained within the Imbrium crater (Deutsch and Stöffler, 1987); this interpretation was devised to support a revisionist view that the Imbrium basin is younger than 3.85 Ga, a concept that is discussed below in secs. 7.5 and 7.6. I believe that remote-sensing data and results from study of Apollo 15 samples have demonstrated conclusively that the Apennine Bench Formation consists of volcanic KREEP basalt flows (Spudis, 1978; Hawke and Head, 1978; Spudis and Hawke, 1986), emplaced after the formation of the Imbrium basin. This early volcanic fill of the Imbrium basin indicates that KREEP volcanism post-dated the Imbrium impact and that this unit is not directly related to the highland deposits of the basin.

At the base of the scarp of the rim of the Imbrium basin are several small isolated patches of fissured and ponded plains that overlie slump masses from the Apennine scarp. These patches of plains have been interpreted as pools of impact melt from the Imbrium basin (Wilhelms *et al.*, 1977; Spudis, 1978; Wilhelms, 1980a), ejected at low velocity from the central cavity and emplaced near the rim of

Figure 7.3 Geological sketch map of Imbrium basin deposits, adapted from Wilhelms and McCauley (1971), Scott *et al.* (1977), Lucchitta (1978), Eggleton (1981), and my own mapping. Secondary crater field omitted for clarity. Base is a Lambert equal-area projection centered on 35° N, 17° W (basin center; see text). After Spudis *et al.* (1988b).

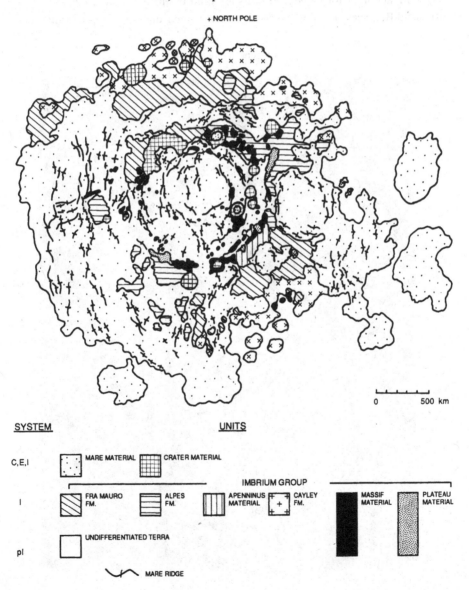

the basin. These isolated melt deposits are not part of the central melt sheet of the basin, which presumably underlies the basalts of Mare Imbrium, but are analogous to the small deposits of melt found within the Montes Rook Formation of the Orientale basin (Chapter 3).

Portions of the ejecta blanket of the Imbrium basin are well exposed (Figures 7.3, 7.4) and permit the detailed subdivision of Imbrium stratigraphy that has been the hallmark of the systematic geological mapping program (summarized in Wilhelms, 1970, 1987). Most of the basin units have been named formally and Spudis *et al.* (1988b) combined these formations into the *Imbrium Group* (Figure 7.4). Formal names for geological units are used in this discussion, but where significant differences occur within or between named units, the formal units have

For legend see page 141.

been subdivided or combined. Most Imbrium deposits appear to be analogous to those of the Orientale basin, but there are differences between the two basins that are not solely attributable to the more advanced degradational state of the Imbrium basin.

The Montes Apenninus range (Figures 7.4a, 7.4b) is a prominent segment of the main topographic ring of the Imbrium basin. Some of the deposits in this range are rough-textured, hummocky materials (AP in Figure 7.4b) that resemble parts of the inner Hevelius Formation of the Orientale basin. This unit has been informally named Apenninus material (Wilhelms and McCauley, 1971) and has a limited distribution around the Imbrium basin, being confined to the southern Apennines (Figure 7.4b). It is composed of rough hummocks on the 10 km scale and some type of ground flow may have been important in its final emplacement. The recognition and mapping of pre-Imbrian craters in the Apennines suggest that this material is 1 to 2 km thick at the Apennine crest (Carr and El-Baz, 1971; Spudis and

For legend see page 141.

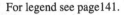

b

Head, 1977). Concentric texture near the crest of the Apennines (Figures 7.4b) suggests that post-basin slumping is partly responsible for the morphology of this unit (Spudis and Head, 1977).

The continuous deposits of the Imbrium basin vary in morphology both concentrically and radially from the basin rim. Those deposits that display a hummocky to radially textured morphology (Figure 7.4a–c) are collectively named the Fra Mauro Formation (Eggleton, 1964; Wilhelms, 1970; Wilhelms and McCauley, 1971). This unit varies widely in thickness and distribution; it occurs mostly northwest and southeast of the basin rim, but is replaced by the Alpes Formation to the northeast and southwest of the basin (Figure 7.3). The Fra Mauro Formation appears to be about 1 km thick near the backslope of the Apennines and thins to a feather edge

For legend see page 141.

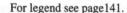

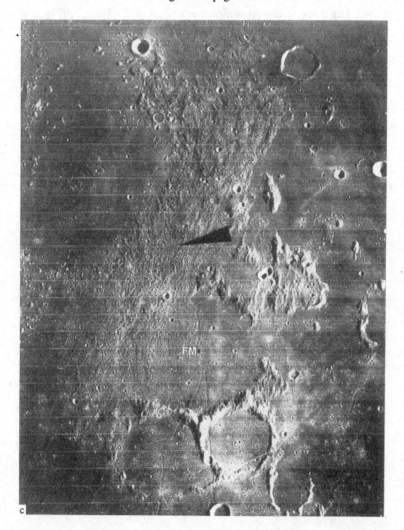

south of the crater Parry, where it interfingers with the Cayley Formation (Figure 7.4c). The Fra Mauro Formation is strongly lineated in the vicinity of the crater Julius Caesar (Wilhelms, 1968) and constitutes part of the "Imbrium sculpture", the radial texture surrounding the Imbrium basin that was first described by Gilbert (1893). The Fra Mauro Formation contains no melt pools as are seen in the Hevelius Formation of the Orientale basin (Chapter 3), or flow lobes, but it frequently appears to have overridden previously formed secondary impact craters and chains (Eggleton, 1970; Head and Hawke, 1975), suggesting that some ground flow may have been important in its final emplacement.

The nature of the Fra Mauro Formation has been the subject of considerable debate. Shoemaker and Hackman (1962), Eggleton (1964), and Wilhelms (1970)

For legend see page 141.

suggested that it is composed of primary ejecta from the Imbrium basin. This view
was challenged on the basis of small-scale impact experiments and theoretical modeling (Morrison and Oberbeck, 1975; Oberbeck, 1975), which indicated that large
amounts of local secondary-crater ejecta are entrained within the deposits collectively known as the Fra Mauro Formation. Such a view was quickly adopted by
other workers studying the geology and samples of the Apollo 14 site (Head and
Hawke, 1975; Hawke and Head, 1977; Simonds *et al.*, 1977b). However, recent
studies of emplacement mechanisms for crater ejecta (Schultz, 1978; Schultz and
Gault, 1985) have suggested that the Fra Mauro Formation may contain much more
primary ejecta than predicted by the Oberbeck (1975) model. This problem will be
addressed below in sections dealing with Apollo 14 site geology (sec. 7.5) and data
synthesis (sec. 7.8).

The next most extensive deposit of the Imbrium basin has been formally named
the Alpes Formation (Page, 1970; Wilhelms, 1970). This unit is a knobby, undulat-

For legend see next page.

e

Figure 7.4 The Imbrium Group, deposits of the Imbrium basin. North at top and framelet width 12 km in each. After Spudis *et al.* (1988b). (a) Northern Apennines, showing massifs (M), the knobby Alpes Formation (AL) and coarsely textured Apenninus material (AP). Apollo 15 landing site shown by arrow. LO IV-102 H1. (b) Southern Apennines. Apenninus material (AP) gradational with hummocky Fra Mauro Formation (F). Dark material (D) is post-basin, pyroclastic mantling deposits. LO IV-109H2. (c) Hummocky Fra Mauro Formation near its type area, the pre-Imbrian crater Fra Mauro (FM; 95 km diameter). Apollo 14 landing site shown by arrow. LO IV-120 H3. (d) Knobby Alpes Formation northeast of Imbrium basin rim. Large crater is Eudoxus (67 km diameter). LO IV-98 H2. (e) Light-plains Cayley Formation (C) in central lunar highlands. Large crater at bottom is Flammarion (75 km diameter); it is cut by gouges (arrows) that make up "Imbrium sculpture" (basin secondary craters) and filled with Cayley Formation. LO IV-108 H3. (f) Imbrium basin secondary craters south of Albategnius. Secondaries form clusters and chains. Floor of Airy (A), a probable basin secondary (Schultz, 1976a), contains cracked, dark material (arrow); this material may be basin melt ejecta. LO IV-101 H2.

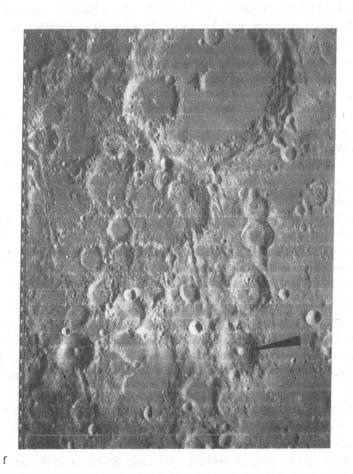

f

ing deposit (Figure 7.4d) and is widespread southwest and northeast of the rim of
the Imbrium basin. The knobs are on the order of 10 km across and are set in a
matrix of rough to smooth highland plains. The Alpes Formation replaces the
expected Fra Mauro Formation in the vicinity of Copernicus, the Aristarchus
Plateau, northern Apennines, and the Plato–Frigoris region. It may extend more
than 800 km from the Imbrium basin rim (Figure 7.3), but appears to be confined
opposite the mare side of both the Montes Alpes and Montes Caucasus fronts, suggesting that these mountains mark the outer bounds of the basin rim. The Alpes
Formation resembles the Montes Rook Formation of the Orientale basin in surface
morphology, but not in distribution. The Montes Rook Formation is confined primarily inside the Orientale basin rim; local occurrence beyond the Cordillera scarp
never exceeds about 50 km (Chapter 3). The characteristic surface morphology of
the Alpes Formation is not easily explainable, but may reflect both the process of
ejecta deposition from the Imbrium basin and surface modification by erosion and
structural deformation. I believe that the surface textures and distribution of the
Alpes Formation indicate that it is ejecta from the Imbrium basin (cf. Head, 1977b).

The unusual bilateral symmetry of the Alpes and Fra Mauro Formations (Figure
7.3) is not seen around any other lunar basin (Spudis, 1986; Spudis *et al.*, 1988b).
(It is possible that the Caloris basin of Mercury displays a similar bilateral symmetry of ejecta facies, but incomplete coverage of the basin by Mariner 10 photogra-
phy prevents direct comparison; see subsec. 9.2.2 and Spudis and Guest, 1988.)
Although it has been proposed that the Imbrium basin is asymmetric in the radial
extent of its ejecta (Wilhelms, 1984, 1987), Figure 7.3 shows that the *radial extent*
of basin deposits is more or less constant; what changes is the *morphological facies*
of ejecta deposits (Spudis, 1986; Spudis *et al.*, 1988b). Moreover, remote-sensing
data (sec. 7.4) suggest that a real, and unexplained, compositional difference exists
between these two geological units. The bilateral symmetry of Imbrium basin
deposits may be a feature of great, if still not understood, significance.

The many small areas of light plains surrounding the Imbrium basin were typi-
cally ascribed a volcanic origin in pre-Apollo studies of lunar geology (Wilhelms,
1970; Milton, 1972; Hodges, 1972). These light plains were formally named the
Cayley Formation, with the type area defined near the distal margins of the
Imbrium basin deposits in the central equatorial region of the Moon near the crater
Cayley (Morris and Wilhelms, 1967). These plains are gradational with the Fra
Mauro Formation towards the Imbrium rim and with basin secondary craters and
chains away from the rim (Figure 7.4c,e,f). The results of the Apollo 16 mission
demonstrated that light plains are of impact origin (see sec. 4.4), and that the
Cayley Formation is somehow related to basin ejecta deposits. In particular, the
gradational nature of the Cayley Formation with large secondary craters suggested
that it is Imbrium basin ejecta, segregated somehow from the continuous Fra Mauro
materials (Eggleton and Schaber, 1972). That the Cayley is composed of primary
Imbrium ejecta was challenged by Oberbeck (1975) and co-workers, who suggested that the process of local mixing is even more significant in the generation of

light plains than it is in Fra-Mauro-type continuous deposits. Some of the Orientale light plains appear to be composed mostly of Orientale ejecta (Chapter 3) and thus, a significant Imbrium-basin-derived component might be expected in the Cayley materials (Wilhelms *et al.*, 1980). Some of the hilly and furrowed materials of the highlands seen elsewhere on the Moon (Howard *et al.*, 1974) may be distal facies of Imbrium ejecta, although it appears that the Descartes materials near the Apollo 16 site are probably derived from the Nectaris basin (Chapter 4). The best example of an isolated deposit that may be primary ejecta from the Imbrium basin is the flow lobe of material filling the floor of the crater Andel M (Moore *et al.*, 1974; Wilhelms, 1980a).

Secondary craters from the Imbrium basin are widely distributed over the Moon (Wilhelms, 1976a, 1980a, 1987; Eggleton, 1981). These craters occur mostly in chains or clusters (Figure 7.4e,f) and are particularly prominent in the central highlands. As discussed previously, Imbrium secondaries may be abundant on the eastern rim of the Serenitatis basin (Chapter 6). Some isolated Imbrium secondaries, such as the crater Airy (Figure 7.4f), display dark, fissured deposits that are probably impact melt ejected from the Imbrium basin (Schultz and Mendenhall, 1979). Chain secondaries may be found as close to the Imbrium rim as Boscovich P, within 200 km of the basin rim and similar in appearance to Vallis Bouvard at Orientale (Chapter 3).

7.3 Imbrium rings and basin structures

The Imbrium basin has one of the largest and most complex ring systems (Figure 7.5) of any lunar basin. Controversy has arisen over the interpretation of the Imbrium ring system and investigators do not agree on their location and fitting of circles to the rings. The first attempt to map the rings of the Imbrium basin in detail was that of Hartmann and Kuiper (1962), who proposed that the Apennine ring is the main topographic rim of the basin; however, this ring is represented not by a topographic high, but by a trough, Mare Frigoris, north of Mare Imbrium. Alternatively, Wilhelms and McCauley (1971) and Howard *et al.* (1974) suggested that the main rim of Imbrium is the northern shore of Mare Frigoris; this reconstruction gives a basin diameter of over 1500 km. On the basis of the distribution of ejecta facies (Figure 7.3), I agree with Dence (1976), who suggested that the large estimates of basin diameter by Wilhelms and McCauley (1971) and Howard *et al.* (1974) are a consequence of attempting to fit the Montes Caucasus into the main ring of the Imbrium basin. Dence (1976) proposed instead that the topographic rim of the Imbrium basin is an arc defined by the Apennines in the southeast and the Sinus Iridum crater in the northwest (Figures 7.1, 7.5). This results in good correspondence to the inner rings of the basin, defined by isolated terra massifs and the mare ridges. The Caucasus range then becomes a structure tangential to the basin, similar to features seen around the Nectaris basin (sec. 4.2). My reconstruction of the ring system of the Imbrium basin results in two interior rings of 550, 790 km diameter and a main rim of 1160 km diameter (Table 2.1; Dence, 1976; Spudis, 1986; Pike and Spudis, 1987).

The innermost Imbrium ring (550 km diameter) is defined by the circular pattern of mare ridges, as seen at other basins (e.g., Crisium, Serenitatis). If any smaller ring exists, it is totally buried by the thick mare basalt fill of the inner Imbrium basin. The intermediate (Archimedes) ring has a diameter of about 790 km and is composed in part of a circular ridge system, isolated massifs such as Mons La Hire, and the Montes Archimedes terra found in the area of the Apennine Bench (Figure 7.1). The main topographic rim of Imbrium, the Apennine ring, is about 1160 km in diameter and is irregular in outline. The northern portion of the ring near the Iridum crater appears to be topographically low, in contrast to the 4 to 5 km elevation of the Apennines. This is not solely attributable to the effects of the impact which formed the Iridum crater (Whitford-Stark, 1981a) and it is probably related to late-stage basin modification processes (sec. 7.8).

It has long been recognized (e.g., Wilhelms and McCauley, 1971; Mason *et al.*, 1976) that the Imbrium basin is situated within a broad pattern of sub-concentric

Figure 7.5 Ring interpretation map of the Imbrium basin, showing six concentric rings; see Table 2.1 for ring diameters. Base is a Lambert equal-area projection centered on 35° N, 17° W (basin center). After Spudis *et al.* (1988b).

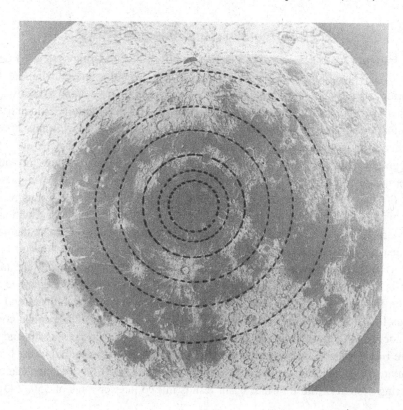

ridges, scarps, and mare shorelines. This regional setting is part of the rationale for postulating the "Procellarum basin" (Whitaker, 1981; Wilhelms, 1987). A consequence of adopting the configuration described here for the main rim of the Imbrium basin (1160 km diameter; centered on 35° N, 17° W) is that the regional concentric pattern is now centered on the Imbrium basin (Figure 7.5; Schultz and Spudis, 1985). Moreover, many of the geochemical arguments advocated for the presence of Procellarum basin (Whitaker, 1981; Wilhelms, 1987) are subject to alternative interpretations (Spudis and Schultz, 1985). For these reasons, I prefer to interpret the regional concentric pattern of the lunar near side as reflecting three large outer rings of the Imbrium basin; Procellarum basin does not exist. These outer rings have diameters of 1700, 2250, and 3200 km (Figure 7.5; Spudis *et al.*, 1988b; Pike and Spudis, 1987). In this interpretation, Imbrium is the largest multi-ring basin on the Moon (Table 2.1).

The Imbrium basin displays tangential structure in the Montes Caucasus and the western segment of the Montes Carpatus (Figures 7.1, 7.3). The Caucasus range consists of platform massifs that display lineations striking radial to the center of the Imbrium basin, but are not perpendicular to the front of this mountain range. These relations could indicate that the Caucasus were in their present position before the final excavation of the central Imbrium cavity. Alternatively, the platform topography of the Caucasus massifs may have been induced by radial structures imposed during basin formation and subsequent platform development (Figure 5.12). Platform massifs also occur in the Apennines (Mons Bradley) and segments of the Montes Carpatus. The Aristarchus Plateau, although lying outside the main Apennine ring, has a rectilinear outline whose boundaries are radial and concentric with an outer ring of the Imbrium basin (1700 km diameter); thus, the Aristarchus plateau may be a platform massif of Imbrium that evolved in response to the extensive volcanism evident in that region (Moore, 1967; Zisk *et al.*, 1977).

Imbrium sculpture (Gilbert, 1893) has been ascribed to both structural (Howard and Masursky, 1968; Scott, 1972a; Schultz, 1976a) and sedimentary causes (Oberbeck, 1975; Head, 1976a). Many elements of Imbrium sculpture are related to aligned secondary craters and chains (Figure 7.4e) and these probably reflect processes of ballistic sedimentation (Oberbeck, 1975). However, several Imbrium radial features are clearly structures, such as the Alpine Valley (Figure 7.1), and contain vents for the eruption of mare basalts, suggesting the presence of deep-seated fractures of structural origin. The Imbrium secondary chain Boscovich P is partly mare-flooded, indicating deep-seated, radial structure associated with the Imbrium basin. Moreover, the location of the lunar caldera Ina (Strain and El-Baz, 1980) on a horst radial to the Imbrium basin suggests volcanism is closely associated with many of these features and that Imbrium sculpture is of both structural and sedimentary origin. This basin-induced radial structure may have affected patterns of lunar structure on a global basis (Figure 7.5; Mason *et al.*, 1976).

The structure of the Imbrium basin is one of vast complexity. Both radial and concentric fractures, which served as vents for the extrusion of later volcanic mate-

Table 7.1 *Chemical composition and mixing model results for Imbrium basin deposits*

Unit	Al wt.%	Mg wt.%	Fe wt.%	Ti wt.%	Th ppm	FAN %	ANGAB %	LKFM %	NOR %	KREEP %	11MB %	12 MB %
1	13.0	5.4	8.6	1.5	3.0	21	–	–	42	9	28	–
2	14.6	4.8	8.6	2.1	7.6	30	–	–	19	30	21	–
3	8.1	5.1	9.9	2.8	5.1	–	7	30	–	28	35	–
4	7.7	5.1	9.1	1.3	9.2	–	5	–	–	82	–	13
5	11.2	4.3	5.8	0.8	5.8	9	41	–	–	50	–	–

Source: Chemical data from La Jolla Consortium (1977), Davis (1980), and Metzger *et al.* (1979); Mixing model results from Hawke and Head (1978) and Spudis *et al.* (1988b).
Note: Regions: 1 – northern Apennines, Alpes Formation (Figure 7.4a); 2 – Southern Apennines, Apenninus material (Figure 7.4b); 3 – combined Apennine Mountains (Figure 7.4a, b); 4 – Fra Mauro Formation (Figure 7.4c); 5 – Ptolemaeus, discontinuous deposits and secondary crater chains (Figure 7.4e).

rials, were produced by the basin impact. These structures may also have undergone long-term modification, producing platform topography, during the epochs of mare volcanism. The Imbrium basin has induced structural patterns over at least an entire hemisphere of the Moon, producing much of the observed topographic and geological complexity evident in the distribution of Imbrium landforms.

7.4 Remote-sensing data: the composition of Imbrium ejecta

Geochemical information is available for portions of the deposits of the Imbrium basin. The extensive mare flooding of the basin interior (Figure 7.1) precludes direct measurements of the chemistry of the melt sheet, but there are data for all other geological units to varying degree. In many highland regions surrounding the Imbrium basin, pyroclastic activity has masked the basin ejecta from direct view (Heiken *et al.*, 1974), and I exclude these areas if possible. Moreover, mare flooding outside the confines of the Imbrium basin has partly buried the Fra Mauro Formation in several areas (Figure 7.3). Thus, only some of the mare components seen in orbital geochemical data may be of Imbrium ejecta origins. Results from studies of lunar samples suggest that mare basalt may be present within the basin ejecta (Ryder and Taylor, 1976). Several mixing models of the various deposits of the Imbrium basin have been done in the Apennines, Fra Mauro region, and the central highlands (Hawke, 1978; Hawke and Head, 1978; Hawke *et al.*, 1980; Spudis, 1986); these results are discussed in this section.

Regional compositions of Imbrium deposits are rich in KREEP (Table 7.1; Spudis *et al.*, 1988b; Metzger *et al.*, 1977; Adler and Trombka, 1977). It appears that anorthositic rocks are not abundant in the Imbrium region; even those areas apparently unmodified by subsequent events display consistently lower Al/Si and higher

Mg/Al ratios than typical highland areas around the Moon (Clark and Hawke, 1981). KREEP has long been associated with the Fra Mauro area (Hubbard *et al.*, 1972) and the Archimedes and Aristarchus regions (Metzger *et al.*, 1973). It appears that some of this KREEP is related to volcanic flows that post-date the Imbrium basin (Spudis, 1978; Hawke and Head, 1978), but Imbrium ejecta are unquestionably enriched in KREEP (Metzger *et al.*, 1979). In the Apennine mountains, the data suggest a more mafic composition in association with KREEP (Table 7.1); such a composition has been interpreted as indicating large amounts of low-K Fra Mauro basalt within the near-rim deposits of the Imbrium basin (Hawke and Head, 1978; Clark and Hawke, 1981; Spudis *et al.*, 1988b). These regional compositions imply that the Imbrium basin excavated materials of deep crustal origin (KREEP and mafic LKFM) and that anorthosites (upper crust) were sparse in the impact target, because either they never formed here or they were removed by older basins.

The Montes Archimedes area (Apennine Bench) is geochemically anomalous; extremely high concentrations of Th (up to 17 ppm) are associated with this terra (Metzger *et al.*, 1979). Such a high Th content suggests that these highlands may consist of KREEP granites and related intrusive rocks (Ryder, 1976) that were emplaced during basin formation. In addition, this material has been embayed by the Apennine Bench Formation of Imbrian age and a significant fraction of the basin deposits probably are masked by these volcanic KREEP basalt flows (Spudis and Hawke, 1986).

A summary of mixing models results for deposits of the Imbrium basin is presented in Table 7.1. The Apennines appear to be dominated by LKFM and KREEP and the anorthositic rocks are subordinate to norite and mare basalt. Th is relatively abundant (Metzger *et al.*, 1979) in the southern Apennines (Apenninus material of Wilhelms and McCauley, 1971), whereas the northern Apennines (dominated by Alpes Formation) show Th levels more akin to those in LKFM basalt. This geochemical difference suggests that these ejecta units are petrologically different and may have been derived from separate stratigraphic levels of the excavated cavity. Mare basalt is also seen in the Apennines, up to 35 percent; this is caused in part by discontinuous dark mantling and partial mare flooding in the Apennine backslope, but some mare basalt probably is present within the basin ejecta.

From 23 near-infrared spectra for the Apennine region, Spudis *et al.* (1988b) distinguished four spectral classes representing KREEP basalts, norites, anorthositic norites, and Mg-gabbronorite (Figure 7.6); distribution of these petrologic units is shown in the map of Figure 7.7. In general, these data are consistent with the orbital chemical data that the bulk of the Imbrium basin deposits in the Apennine region is noritic. Several areas (class 4, Figure 7.7) are made up of Mg-gabbronorite, a highland rock type composed of both clinopyroxene and orthopyroxene (James and Flohr, 1983). Mg-gabbronorites appear to have formed from separate magmas intruded into the crust after the global magma ocean solidified (Wood *et al.*, 1970; Warren, 1985); their presence in the deposits of the Imbrium basin attests to a long and complex igneous history at the impact site.

Figure 7.6 Representative spectra of the four classes measured by Spudis *et al.*
(1988b) for geological units in the Apennine Mountains of the Imbrium basin.
Class 1 = norite; Class 2 = anorthositic norite; Class 3 = KREEP basalt; Class 4
= Mg-gabbronorites. After Spudis *et al.* (1988b).

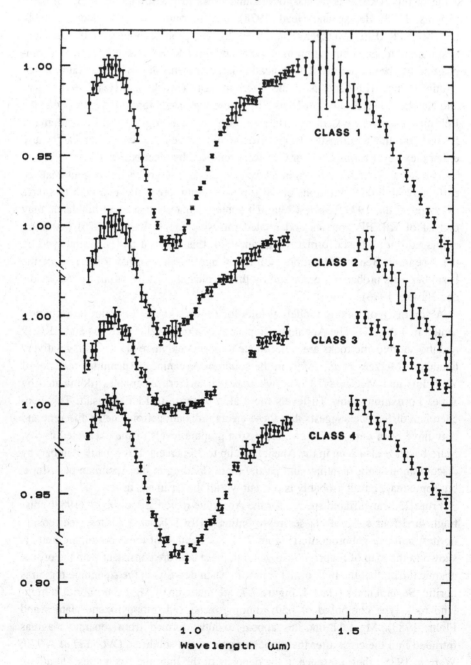

The compositions of the Fra Mauro region are dominated by KREEP for the highlands and mare basalt for the plains that embay them (Table 7.1; Hawke and Head, 1978). This regional petrologic makeup is consistent with the samples from the Apollo 14 site (discussed below,) most of which are KREEP-rich, basaltic breccias. Anorthositic rocks make up only about 5 percent of this region. The central highlands east of Fra Mauro crater show decreasing KREEP content and an increase in anorthositic components (Table 7.1). Anorthositic rocks and KREEP are present in subequal amounts in this area. The differences in composition between the Fra Mauro and central highlands areas may be caused by the discontinuous distribution of the Fra Mauro Formation in the latter. The mixing models for the central highlands (Tables 4.1, 7.1) suggest that the regional compositions are controlled by local bedrock geology and that the KREEP component was deposited only discontinuously by the Imbrium ejecta. The effects of local, non-basin events on the formation of some of these regional compositions are further demonstrated by the composition of the Aristarchus Plateau (Zisk *et al.*, 1977), where KREEP volcanism of Imbrian age has masked the local composition of ejecta from the Imbrium basin.

Figure 7.7 Index photograph showing locations of the four spectral classes of Spudis *et al.* (1988b). See Figure 7.6 for representative spectra and rock interpretations.

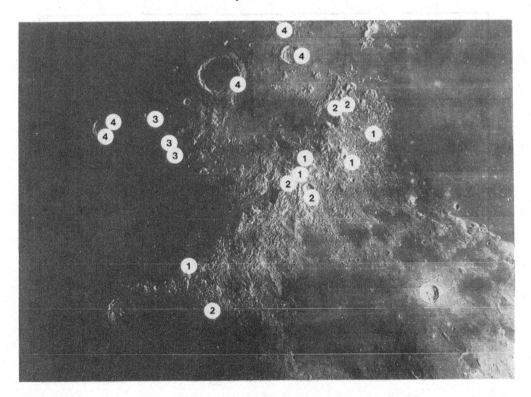

The remote-sensing data have several important implications for the geology of the Imbrium basin. The impact target for the Imbrium basin did not possess the significant layer of anorthosite, as is present at the Orientale, Nectaris, and Crisium basins. Imbrium ejecta are dominated by KREEP basalts, norites, low-K Fra Mauro basalts, and mare basalts. Minor quantities of intrusive, KREEP-rich rocks (granites and quartz monzodiorites), Mg-gabbronorites, and mafic rocks of the Mg-suite, such as dunite, were emplaced within the main basin rim and possibly make up a small fraction of the continuous ejecta. Material derived from lower levels of the crust (LKFM basalt) is found mostly within the Apennines, suggesting that these were the deepest crustal layers excavated. No evidence is seen for ultramafic material from the lunar mantle, suggesting that if such material was excavated by Imbrium, it may be confined within the inner basin and covered by subsequent flows of mare basalt. The impact target for the Imbrium basin may be characterized as a KREEP-rich, noritic igneous complex, undergoing active KREEP and mare volcanism at the time

Figure 7.8 Geologic cross section of the Apollo 14 landing site, Fra Mauro. Fra Mauro Formation at site is a complex mixture of Imbrium basin primary ejecta and local materials; the exact proportions of each remain unknown. Cone crater excavated materials of the ridged member, here interpreted as Fra Mauro breccias and as primary ejecta from the Imbrium basin. By the author from Heiken *et al.* (1991).

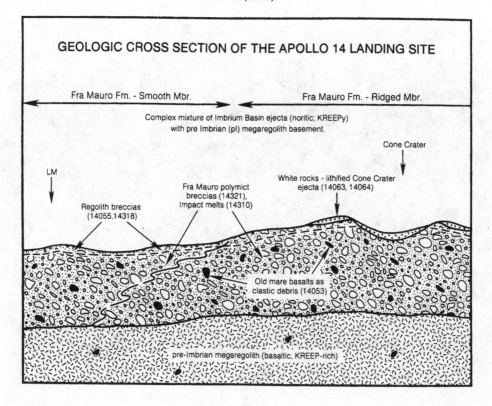

GEOLOGIC CROSS SECTION OF THE APOLLO 14 LANDING SITE

Fra Mauro Fm. - Smooth Mbr.   Fra Mauro Fm. - Ridged Mbr.

Complex mixture of Imbrium Basin ejecta (noritic; KREEPy)
with pre Imbrian (pI) megaregolith basement.

Cone Crater

LM

White rocks - lithified Cone Crater
ejecta (14063, 14064)

Fra Mauro polymict
breccias (14321),
Impact melts (14310)

Regolith breccias
(14055,14318)

Old mare basalts as
clastic debris (14053)

pre-Imbrian megaregolith (basaltic, KREEP-rich)

of basin formation. Thus, the Imbrium target is similar to that inferred for the Serenitatis basin, with the exception that the Imbrium site contained some anorthosites and greater KREEP. These petrologic characteristics are consistent with the complex history for this region inferred from the photogeological evidence.

7.5 The Fra Mauro Formation: petrology of the Apollo 14 site

The Apollo 14 landing site is located within the continuous deposits of the Imbrium basin (Figure 7.4c), near the type area for the Fra Mauro Formation (Swann *et al.*, 1977). The spacecraft landed near the fresh crater Cone, which was presumed to have excavated material from a ridge of Fra Mauro material below the local regolith (Figure 7.8). This ridge has been interpreted as an accumulation of primary ejecta from the Imbrium basin (Swann *et al.*, 1977), or alternatively, as the rim of a large, pre-Imbrian crater (Hawke and Head, 1977b). The debate over the proportions of locally derived and primary ejecta from Imbrium in the Fra Mauro Formation hinges in part on this interpretation. Abundant secondary craters of the Imbrium basin appear to be buried by the Fra Mauro Formation in this region (Eggleton, 1970; Head and Hawke, 1975). Thus, mixing of primary ejecta with secondary-crater ejecta did occur (Oberbeck, 1975) and we must determine which Apollo 14 samples are Imbrium primary ejecta and which are derived locally.

The most important samples from the Apollo 14 site are those collected at the rim of Cone crater, which presumably represents the local Fra Mauro Formation "bedrock". The rocks are breccias with crystalline matrices and an extremely high clast content (Figure 7.9); they have been collectively named Fra Mauro breccias (Wilshire and Jackson, 1972). They have a basaltic bulk composition, are rich in KREEP, and contain clasts of impact melts, anorthositic gabbro, mare basalts, granites, and other breccias (Wilshire and Jackson, 1972; Grieve *et al.*, 1975; Simonds *et al.*, 1977). The breccias themselves are impact melts (Simonds *et al.*, 1977) and appear to have undergone multiple episodes of brecciation (Wilshire and Jackson, 1972; Grieve *et al.*, 1975). The best example of the Fra Mauro breccias is 14321 (Duncan *et al.*, 1975; Grieve *et al.*, 1975). This rock has a bulk chemical composition of basalt, with KREEP trace-elements, abundant clasts, and breccia-within-breccia texture. The clast populations of the Fra Mauro breccias, dominated by basalts and reworked breccia clasts, suggest near-surface derivation, as would be expected for ejecta at this range from the Imbrium basin.

I interpret the Fra Mauro breccias as clastic, primary ejecta from the Imbrium basin, as suggested by Wilhelms *et al.* (1980). Wilhelms (1987), acknowledging the origin of the matrices of the Fra Mauro breccias as impact melt (Simonds *et al.*, 1977), suggested that this melt is from the melt sheet produced by the Imbrium basin. I disagree with this interpretation; in my view, the impact melt that makes up the matrices of the Fra Mauro breccias, claimed as evidence for a local origin by Simonds *et al.* (1977), was produced near the surface of the target site of the Imbrium basin by heavy cratering and basin formation in pre-Imbrian time. The breccias were then transported to Fra Mauro as clastic ejecta during the formation of the Imbrium basin.

Other rocks collected at the Apollo 14 site have petrography and chemistry consistent with local origins. Regolith breccias contain Fra Mauro breccia fragments as clasts (Wilshire and Jackson, 1972); these rocks were produced by regolith formation on the Fra Mauro Formation. A group of poorly lithified breccias, informally called the "White Rocks" (Marvin, 1976; Ryder and Bower, 1976), are highly shocked, fragmental breccias that may be analogous to the feldspathic fragmental breccias collected at North Ray crater at the Apollo 16 site (sec. 4.4). If they are so analogous, the white rocks are the lithified products of the impact that produced Cone crater. Considerable attention has been drawn to a crystalline rock, 14310 (James, 1973; Swann *et al.*, 1977). This sample is composed of nearly pure KREEP basalt and was originally thought to be of volcanic origin (Hubbard *et al.*, 1972). It has since been shown that it is an impact melt (James, 1973) and has a composition similar to local components found near the site (Schonfeld and Meyer, 1972). Thus, 14310 probably was produced by a local impact event that remelted materials found within the Fra Mauro Formation.

Another interpretation of the Fra Mauro breccias and their geological setting has

Figure 7.9 Texture of 14314,12, a Fra Mauro breccia. Rock has a crystalline matrix and shows evidence for multiple brecciation and contains shallow level clastic debris, such as mare basalts. Field of view 3 mm, plane light.

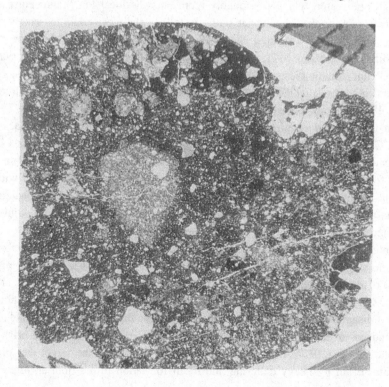

significant implications for the age of the Imbrium basin. Deutsch and Stöffler (1987) and Stadermann *et al.* (1991) contend that the Fra Mauro breccias represent a variety of impact melt and breccia formations from many local impacts, some of which were then incorporated into a "megabreccia" deposit by the "debris surge" from the emplacement of Imbrium ejecta, as suggested by the model of Oberbeck (1975). Thus, clasts of impact melt contained within this megabreccia should pre-date the formation of the Imbrium basin and the Imbrium impact can be no older than the age of the youngest sample of impact melt (Deutsch and Stöffler, 1987). Radiometric dating of a variety of clasts from the Apollo 14 and 16 sites suggested to Stadermann *et al.* (1991) that the Imbrium basin is about 3.75 Ga old, not the 3.85 Ga advocated by Wilhelms (1987) and myself in this book.

I reject these hypotheses for a variety of reasons. All of the melt clasts dated by Stadermann *et al.* (1991) are from the Cone crater at the Apollo 14 site, including the "white rocks", which are lithified products of the impact which formed Cone crater. While some of these samples are probably from the Fra Mauro Formation, others are discrete samples, untraceable, and perhaps unrelated, to the Fra Mauro Formation. Nearly all of the young impact melts studied by this group are of basaltic, KREEP-rich composition (Table 2 of Stadermann *et al.*, 1991), comparable to the gross composition of the Fra Mauro Formation and consistent with their formation by local impacts after emplacement of the unit (i.e., after formation of the Imbrium basin). The arguments given by Deutsch and Stöffler (1987) against an origin by local, post-basin impact based on the petrographic textures of these samples are equivocal and model-dependent. We have no experience with the thermal histories of impact melt sheets, and consequent petrographic textures, for crater sizes of hundreds of meters to several kilometers in diameter. Moreover, one sample of Cone crater ejecta, 14063, has no clast of impact melt younger than 3.86 Ga (Fig. 5 of Stadermann *et al.*, 1991), exactly the relation expected for the Fra Mauro Formation if the Imbrium basin were 3.85 Ga old. I accept the 3.85 Ga age of the Imbrium basin mostly on the basis of the ages of Apollo 15 samples, especially the well determined age of Apollo 15 KREEP basalt (discussed below in sec. 7.6).

The Fra Mauro breccias collected at the Apollo 14 site are clastic primary ejecta from the Imbrium basin. The impact melt, mare basalt, and older breccia clasts contained within the Fra Mauro breccias indicate that the target site for the Imbrium basin was basaltic in composition and KREEP-rich; it may have consisted of highlands that were resurfaced by KREEP and mare basalt lavas in pre-Imbrian time. These data confirm the inferences made on the basis of mixing models of orbital data that the target site for the Imbrium basin was complex and of atypical composition, compared to other sites in the lunar highlands.

7.6 The Apennine Mountains: petrology of the Apollo 15 site

The mission to the Hadley-Apennine site returned samples and data on the main topographic rim of the Imbrium basin. The objectives of Apollo 15 were to characterize the nature of the Apennine front, directly sample Imbrium ejecta, and to return samples of pre-Imbrium-basin rocks that might be exposed within massifs at the site (Carr and El-Baz, 1971; Howard, 1971). The data from the Apollo 15 mission provide important constraints on the geology of the Imbrium basin (Spudis, 1980; Spudis and Ryder, 1985, 1986).

Although geological control is poor, rocks related to the Imbrium basin apparently were returned by the Apollo 15 mission. The height of the Apennine front near the Apollo 15 site (Figure 7.10) is more than 3400 meters; this relief is in excess of the estimated thickness of Imbrium ejecta near the Apennine crest (1000–2000 meters; Carr and El-Baz, 1971). Thus, a large section of pre-Imbrian rocks should be exposed within the front. The highlands at the site were sampled at Hadley Delta (Figure 7.10), a large, equant massif that appears to have a relatively

Figure 7.10 Detail of the Apollo 15 landing site, Apennine Mountains. Landing point was on mare basalt flows; Apennines were sampled at massif Hadley Delta. Apennine Bench Formation underlies mare basalts at site; KREEP basalts returned by Apollo 15 are derived from this unit. Dark mantle deposits consist of pyroclastic glasses. North at top, field of view about 150 km. Part of AS15-M-415. From Spudis and Ryder (1985).

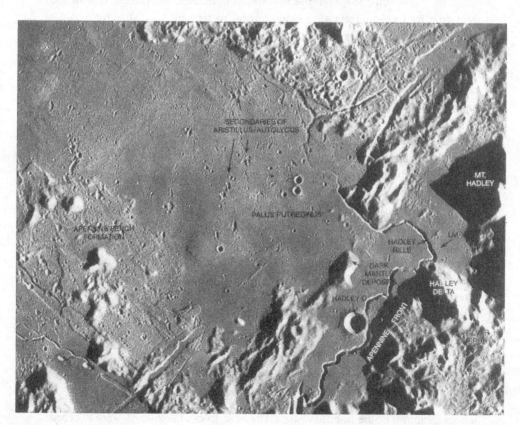

Table 7.2 *Composition of melt rocks and glasses from the Apennine Mountains,
Apollo 15 landing site*

	1[a]	2[b]	3[c]	4[d]	5[e]	6[f]
TiO_2	1.47	1.35	1.08	1.31	1.25	1.45
Al_2O_3	16.66	17.1	17.9	16.1	18.83	18.01
FeO	9.83	8.79	8.6	8.6	9.67	8.94
MgO	16.0	13.3	12.7	13.8	11.0	12.41
K_2O	0.14	0.17	0.73	0.35	0.12	0.21
Mg^*	72.2	75.6	72.4	74	66.8	71

Notes: [a]Station 7 boulder, black and white rock 15445, aphanitic melt breccia (Ridley *et al.*,
1973)

[b]Loose sample, black and white rock 15455, aphanitic melt breccia (S.R. Taylor *et al.*,
1973).

[c]Rake sample 15359, poikilitic impact melt breccia, group C (Ryder and Spudis, 1987).

[d]Coarse fines particle 15314,146, fine-grained intersertal melt breccia, group B (Ryder and
Spudis, 1987)

[e]Average of 82 analyses of LKFM impact glasses from Apennine Front, Apollo 15 (Reid *et
al.*, 1972).

[f]Apollo 17 poikilitic melt sheet (Table 6.2)

$Mg^* = (Mg / Mg+Fe)$ (atomic)

mature surface; few large boulders are exposed on its slopes, in contrast to the massifs of the Taurus–Littrow valley near the Apollo 17 site (Chapter 6).

For understanding basin geology, the most important Apollo 15 samples are the
impact melt rocks (Ryder and Spudis, 1987). One of the most interesting varieties
of melt rocks returned from the Apollo 15 site are the so-called "black-and-white"
rocks, 15445 and 15455 (Ridley *et al.*, 1973; Ryder and Bower, 1977; Ryder and
Wood, 1977). These rocks are aphanitic impact melt breccias that are relatively
clast-rich (Figure 7.11). The chemistry of the melt phase is Mg-rich, low-K Fra
Mauro basalt, considerably more magnesian than comparable poikilitic melts from
Apollo 17 (Table 7.2). The Apollo 15 melts also contain clasts of deep crustal or
upper mantle origin (Ryder and Bower, 1977; Herzberg, 1978) and such clasts
include pristine norites, troctolites and spinel troctolites. Based in part on analogy
to the Apollo 17 melt rocks, the Apollo 15 black-and-white rocks have been interpreted as fragments of the impact melt sheet of the Imbrium basin (Ryder and
Bower, 1977; Ryder and Wood, 1977).

Several other impact melts, found as small fragments in the Front soils, were
collected from the Apennine Front, but their provenance is uncertain (Figure 7.12;
Ryder and Spudis, 1987). It is possible that some of these other melt groups, and
not the black-and-white rocks, represent the Imbrium melt sheet. The most likely
group for such an origin are the rocks of melt group "B" of Ryder and Spudis
(1987); this is the most populous group of impact melt rocks and they have a bulk

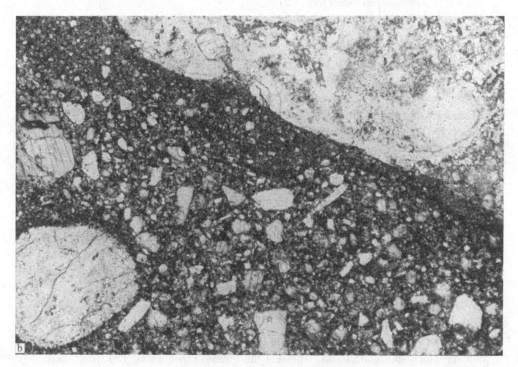

composition of LKFM, inferred to be indicative of basin impact melt (Spudis, 1984; Spudis *et al.*, 1991). None of these melts are large rocks, but many have been dated radiometrically; those with good plateaux have ages around 3.87 Ga (Dalrymple and Ryder, 1991).

On the basis of clast contents and bulk composition, Ryder and Spudis (1987) considered that the Apollo 15 black-and-white rocks 15445 and 15455 (dated at 3.85 Ga; Alexander and Kahl, 1974) are the best candidates for samples of the melt sheet of the Imbrium basin. They are more mafic than the Apollo 17 poikilitic melts, reflecting origin deeper within the Moon, presumably because the impact forming the Imbrium basin was much larger than the one that formed Serenitatis. On the basis of observed chemistry and petrology of the Apollo 15 melts as well as the observation of Imbrium basin melt ponds that occur along the margins of the Apennine front, I believe that 15445 and 15455 are fragments of ejected melt of the Imbrium basin, although future study of other melt rocks from the Apollo 15 landing site may reveal more likely candidates.

Other rocks returned from the Apennines appear to be of non-Imbrium basin derivation. Anorthosite is sparse at the Apennine front, in accord with its low abundance inferred from the remote-sensing data (sec. 7.4). The large sample 15415 (the "Genesis rock") is a ferroan anorthosite (James, 1972, 1980) and has a pre-Imbrian metamorphic age of 4.15 Ga (Husain *et al.*, 1972). This rock is probably clastic ejecta from the Imbrium basin or is related to the pre-Imbrian section exposed within the front, described below. Many regolith breccias were returned from the highlands stations at Apollo 15 and contain clasts of KREEP basalt, local mare basalts, and glass (Dymek *et al.*, 1974; Ryder and Bower, 1977); these rocks are products of regolith development on the Hadley Delta massif, similar to the Apollo 14 regolith breccias (sec. 7.5).

Abundant samples of pristine, volcanic KREEP basalts are present on Hadley Delta (Dowty *et al.*, 1976; Irving, 1977; Ryder, 1987) and probably are derived from the Apennine Bench Formation; therefore, they are not products of the impact that formed the Imbrium basin (Spudis, 1978, 1980; Hawke and Head, 1978; Spudis and Ryder, 1985). Deutsch and Stöffler (1987), while not questioning the volcanic origin for the KREEP basalt, argue that (1) the KREEP basalt is not derived from the Apennine Bench Formation, and (2) even if it is so derived, the Apennine Bench plains are not post-Imbrium-basin volcanic flows, but a "megablock" contained within the basin cavity. Spudis and Hawke (1986) demon-

---

Figure 7.11 Apollo 15 aphanitic impact melt 15455, one of the "black-and-white rocks", interpreted by Ryder and Bower (1977) as fragments of the melt sheet from the Imbrium basin; I interpret these rocks as ejected melt from the Imbrium basin. (a) Sample 15455 from Spur crater in the Lunar Receiving Laboratory. Black matrix is the melt phase; the large white clast is a pristine norite. (b) Photomicrograph in transmitted plane light; field of view about 2.5 mm. Matrix is extremely fine-grained and contains abundant clastic mineral debris.

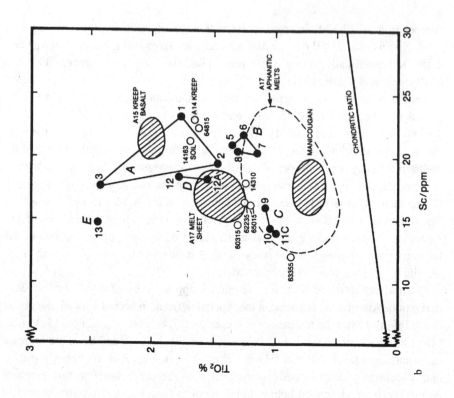

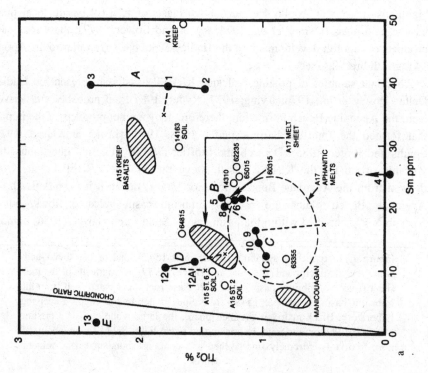

strated that the composition of the Apennine Bench Formation and Apollo 15 KREEP basalt is identical, chemically and mineralogically; the claim by Deutsch and Stöffler (1987) that the MgO contents for the two do not match is incorrect, taking into account that the error estimate for the orbital data (MgO = 5.7 ± 2.0 wt.%, inadvertently omitted from the table in Spudis and Hawke, 1986) encompasses MgO values for Apollo 15 KREEP basalt (MgO = 6.3–8.2 wt.%; see Irving, 1977 and Meyer, 1977). Moreover, photogeological evidence clearly shows that the Apennine Bench plains embay (and therefore post-date) terra of the Imbrium basin (Hackman, 1966; Wilhelms, 1970, 1987; Wilhelms and McCauley, 1971; Hawke and Head, 1978; Spudis, 1978); the plains are *not* a structurally emplaced megablock. Some of the Apollo 15 KREEP basalts might be Imbrium ejecta because the orbital data suggest that local areas of high Th concentration occur within the Apennines (Metzger *et al.*, 1979), which are ejecta from Imbrium. However, most KREEP basalts found at the landing site are derived from the Apennine Bench Formation, an exposure of volcanic flows emplaced after the formation of the Imbrium basin. This interpretation is a significant constraint on the age of the Imbrium impact; because of the embayment of Imbrium terra by post-basin KREEP basalt, the basin can be no younger than the crystallization age of these basalts, which cluster closely around 3.85 ± 0.05 Ga (e.g., Nyquist *et al.*, 1975; Carlson and Lugmair, 1979).

Low-K Fra Mauro basalt is the dominant chemical component in Apennine front soils (Warner *et al.*, 1972b); in particular, impact glasses of LKFM basalt composition are abundant. These glasses have varied bulk chemistries, reflecting mixing during regolith development (Reid *et al.*, 1972), but are consistently less magnesian than the LKFM matrices of the black-and-white rocks (Table 7.2). These glasses more closely resemble the Apollo 17 poikilitic melt rocks, which are derived from the Serenitatis basin (Chapter 6), than they do the probable Imbrium melt rocks 15445 and 15455. The presence of such a component within the Apennine Front suggests that a large section of Serenitatis basin ejecta is exposed at the Apollo 15 landing site (Figure 7.13); such a relation was predicted on the basis of pre-mission photogeological analysis (Carr and El-Baz, 1971). An unusual ultramafic green glass, found mainly in Apennine front soils, was postulated to be impact melt from Imbrium, derived from the lunar mantle (Dence *et al.*, 1974; Dence, 1977a). Recent study, however, suggests that its chemistry is compatible with a volcanic origin

Figure 7.12 Chemical variation in Apollo 15 impact melts. (a) Variation in Ti and Sm. Apollo 15 melts form five separate groups (A–E), each one distinct from impact melts from the Apollo 16 (Nectaris basin) and Apollo 17 (Serenitatis basin) groups. Melt sheet of terrestrial Manicouagan crater (Floran *et al.*, 1978) shown for reference. (b) Variation in Ti and Sc. Clustering of Apollo 15 melt compositions mimics pattern seen in (a). Each group appears to be distinct from melt rocks at other lunar highland sites. From Ryder and Spudis (1987).

(Delano, 1979), analogous to the Apollo 17 pyroclastic glasses (Heiken *et al.*, 1974).

The Apollo 15 mission apparently was successful in sampling both Imbrium basin ejecta and pre-Imbrian rocks (Figure 7.13). The samples of Imbrium impact melt extend the data base for impact melts from lunar basins and provide information on the composition of the lower crust in the Imbrium region, much as the Apollo 14 samples provide data on the composition of shallow levels of the basin target. Moreover, a section of pre-Imbrian rocks within the Apennines supports the postulated structural origin for the Apennine ring (sec. 7.8). This section exposes mostly Serenitatis basin ejecta, although some pre-Serenitatis material may be represented by extremely old rocks from the site (e.g., 15415). Finally, through both samples of impact melt, some of which probably formed in the basin impact, and the sampling of a volcanic unit which was extruded shortly after the formation of the basin (the Apennine Bench Formation), we have a good estimate of the age of the Imbrium basin, a key event in lunar geological history: this impact occurred 3.85 Ga ago. Thus, Apollo 15 may be the key to the unraveling of the geology of the Imbrium basin and the results of this mission constrain basin development.

Figure 7.13 Geologic cross section summarizing relations between units at the Apollo 15 landing site. By the author in Heiken *et al.* (1991).

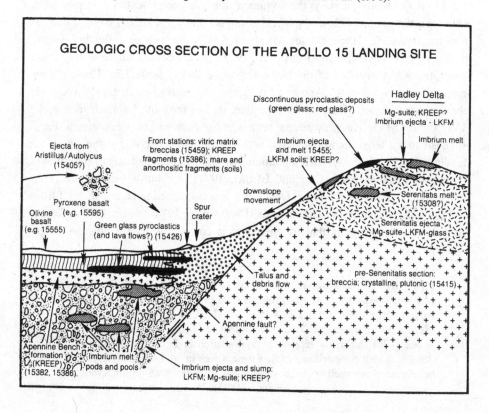

7.7 Petrology of the Apollo 16 site - Imbrium basin ejecta?

The notion persists that the samples returned from the Apollo 16 landing site may be mostly derived from the Imbrium basin (Eggleton and Schaber, 1972; Wilhelms *et al.*, 1980; Hodges and Muehlberger, 1981). As discussed in Chapter 4, there are compelling reasons to believe that most Apollo 16 samples are derived from the Nectaris basin and some samples are from local craters. If material from the Imbrium basin exists in the Apollo 16 sample collection, it has not been recognized to date. Wetherill (1981) postulates that Apollo 16 KREEP (meaning LKFM poikilitic impact melts; sec. 4.4) is melt ejecta from the Imbrium basin; this interpretation is based on the assumption that KREEP is found on the Moon only in the Imbrium region. This assumption is incorrect (Hawke and Head, 1978; Spudis, 1979; Hawke and Spudis, 1980); KREEP deposits are found in many highland areas, including the lunar far side (Spudis, 1979). Moreover, the LKFM within Nectaris basin deposits (Chapter 4) suggests that LKFM existed at the target site of the Nectaris basin (Maurer *et al.*, 1978; Spudis, 1984) and that an Imbrium origin for LKFM found at the Apollo 16 site is not required, although such an origin is not excluded (Spudis, 1984).

There is little to constrain estimates of exotic contributions from the Imbrium basin to the Apollo 16 samples. Anorthosite found at Apollo 16 in abundance is sparse to absent at the Imbrium-related Apollo 14 and 15 sites; moreover, the Apollo 14 anorthosites are more alkalic than the ferroan Apollo 16 anorthosites (Warren and Taylor, 1981). These samples at Apollo 16 probably are derived locally, or at least, come from regions near the site (e.g., the Nectaris basin). Although Imbrium ejecta may exist at the Apollo 16 site, there are at present no distinguishing criteria for their recognition.

7.8 The formation and evolution of the Imbrium basin

The Imbrium basin formed in a relatively thin crust of extraordinary geochemical and petrological complexity. Many older basins extensively modified, and may have partly removed, the pristine lunar crust in the target area. It appears that ferroan anorthosite, if ever present in this region, mostly was removed by these large impacts. The Imbrium target had been the site of intense igneous activity, both intrusive and extrusive. Copious mafic plutonism is represented by the Mg-suite rocks in the Apennine region (sec. 7.4) and by abundant clasts of plutonic rocks contained within the Apollo 15 black-and-white breccias. Extensive volcanism is inferred from the variety of mare basalt clasts contained within basin ejecta collected at Apollo 14, as well as global evidence for pre-Imbrian mare (Schultz and Spudis, 1979) and KREEP volcanism (Schonfeld and Meyer, 1973; Hawke and Spudis, 1980). KREEP plutonism may be responsible for the production of KREEP-rich granitic rocks returned by the Apollo 15 mission (Ryder, 1976) and evident in the orbital data as components of the inner basin deposits (Metzger *et al.*, 1979).

The lunar lithosphere was probably fairly thick at the time of the Imbrium impact (Ferrari *et al.*, 1978; Solomon and Head, 1980), although it may have been at least partly penetrated during basin formation. This penetration is inferred from the platform topography evident in the Montes Caucasus and Carpatus and the extensive volcanism associated with the structurally controlled Aristarchus Plateau (sec. 7.3). The main basin rim appears to be missing in the Sinus Iridum region, suggesting that sub-lithospheric flow after basin formation was important in the development of that topography (Whitford-Stark, 1981a). Alternatively, the rim topography in this region may have been suppressed by interaction with older basin rings.

The pre-Imbrian Insularum and Serenitatis basins had profound effects on the development of the topography of the Imbrium basin. The prominent relief of the Apennines is mostly caused by its location outside the main rims of both of these two basins. The development of the Apennine Bench is probably related to the intersection of the Apennine scarp with the outer ring of the older Insularum basin (Spudis and Head, 1977; Head, 1977b). Coincident rings of Imbrium and Serenitatis accentuated the Apennine topography whereas the discordant rings of Insularum in the Montes Carpatus suppressed the topographic prominence of this segment of the Imbrium basin rim. In addition, the Montes Carpatus has platform morphology, suggesting that sub-lithospheric flow may have been an important process in this part of the basin rim.

The transient cavity of the Imbrium basin must lie inside the intermediate (Archimedes) ring in order to preserve the pre-Imbrian, Apennine Bench which lies within the basin rim. Thus, the maximum diameter for the transient cavity is about 800 km. The minimum value is more uncertain; analogy with other lunar basins suggests that the diameter of the inner ring (about 600 km) probably represents an adequate lower bound for this value. All of the ejecta in the Imbrium deposits are of crustal origin; material from the lunar mantle has not been recognized. The dearth of mantle material in the ejecta suggests that the depth of basin excavation was probably little more than the total thickness of the crust, here about 50 to 60 km. The model of Croft (1981a) predicts excavation depths of 60 to 85 km, on the basis of estimates of transient cavity diameter. I consider this reasonably good agreement, because much of the Imbrium basin ejecta, particularly the near-rim deposits, are buried by mare basalts. The lunar mantle may have been excavated by the impact which formed the Imbrium basin, but such material has not been identified.

Modification of the Imbrium basin after the excavation phase was extensive. The outer Apennine, and possibly the Archimedes, rings were formed by large-scale slumping inward toward the basin center immediately after the impact, producing megaterraces (Mackin, 1969; Dence, 1976, 1977a; Head, 1977a; Spudis and Head, 1977; Croft, 1981b). Megaterrace formation was responsible for generating the unusual topography of Imbrium by interaction with older basin rings, resulting in topographic anomalies such as the Apennine Bench. Sub-lithospheric flow in the northern portion of the basin suppressed the development of the outer ring near the Sinus Iridum crater and possibly, in the Montes Alpes.

The Imbrium basin displays prominent tangential structure in the Montes Caucasus and Carpatus (Figure 7.1), suggesting that structural modification to the basin occurred well outside its transient cavity. The morphology of the Montes Caucasus suggests that this range may have been emplaced before excavation of the central cavity was complete (sec. 7.3); structural modification continued long after the Imbrium impact, illustrated by the development of platforms in these areas (Figure 7.1) and by extensive post-basin volcanism in the Aristarchus Plateau. Moreover, the occurrence of a platform massif within the Apennines (Mons Bradley) indicates that such structural adjustment was a basin-wide phenomenon and not solely limited to the massif ranges tangential to the basin rim. Radial structures served as loci for volcanic extrusion; the almost immediate emplacement of the Apennine Bench Formation after Imbrium formed (Spudis, 1978) suggests that volcanism was active during this extended time of topographic modification.

Ejecta of the Imbrium basin were sampled in the Apennines and at Fra Mauro and come from two distinct crustal levels. The near-rim Apennine site provided samples of the lower crust. These samples are low-K Fra Mauro basalt and mafic cumulates of the Mg-suite. The distal portions of Imbrium ejecta (the Fra Mauro Formation) are composed of KREEP and mare basalts, extensively reworked by basin and crater impacts that occurred before the Imbrium basin formed. These relations are consistent with a surficial, or at least upper crustal, origin for the Fra Mauro Formation at the Apollo 14 landing site. Anorthositic rocks are rare in ejecta from the Imbrium basin, suggesting that anorthosite was not a common rock type in the impact site. The Apollo 16 samples are mostly Nectaris ejecta and the postulated component of Imbrium ejecta (e.g., Hodges and Muehlberger, 1981) has not been identified.

Results from remote-sensing and sample data indicate that the target site for the Imbrium basin was extremely heterogeneous on a regional scale. Moreover, basin ejecta may be extensively mixed, with materials from the lower crust juxtaposed with upper crustal rocks at the same radial distance from the basin, as suggested by the mixture of the deeply derived LKFM and upper crustal anorthosite within the Apennines (sec. 7.4). Such a relationship is seen in some ejecta blankets of terrestrial craters (Hörz and Banholzer, 1980). The data further support the idea that the lunar highlands are composed of geochemical provinces (Spudis and Hawke, 1981) that have distinct geological and petrological histories; the target for the Imbrium basin is different from any of the other basins on the Moon.

Whereas the Imbrium basin resembles many of the basins described in this book in some aspects, its geological development is unique. This is a consequence of both its large size (the largest basin on the Moon) and its time of formation, when the Moon's thermal state was in transition from an early, thin to a late, thick lithosphere. Thus, the Imbrium basin possesses some attributes of older style basins, such as Crisium, in the development of platform massifs and internal modification.

However, Imbrium also is analogous to young basins, such as Orientale, in the interaction of its outer structures with older landforms, as well as in its comparable ejecta morphologies and depositional processes. The Imbrium basin is a transitional feature in many respects and provides the last link in the geological model for the formation of lunar basins developed in the next chapter.

8

Geological processes in the formation of lunar basins

The formation of multi-ring basins dominated the early geological evolution of the Moon. The five basins described in the preceding chapters represent a spectrum of basin ages, sizes and morphologies. By comparing the similarities and differences among these basins, some general inferences may be made regarding the process of formation of multi-ring basins on the Moon. I here synthesize the information described in the previous chapters to develop a model for the formation and geological evolution of multi-ring basins on the Moon. This model is incomplete, but several puzzling aspects of basin geology can be explained satisfactorily through this approach. At various points in the following discussion, please refer to preceding sections in the text.

8.1 Composition and structure of the lunar crust

The crust of the Moon is heterogeneous on a local and a regional scale; the impact targets for lunar multi-ring basins were similarly heterogeneous. The crustal thickness at the basin target sites was widely varied, ranging from 50 km thick for parts of the Imbrium basin to over 120 km thick for the Orientale highlands (Bills and Ferrari, 1976). Moreover, lithospheric conditions during the era of basin-forming impacts changed with time in response to rapidly changing thermal conditions within the Moon 4 Ga ago (Hubbard and Minear, 1975; Solomon and Head, 1980). The older basins formed in a relatively thin, easily penetrated lithosphere that gave rise to extensive post-impact modification. Young basins (e.g., Orientale) were formed after the Moon's outer layer had become rigid, and opportunities for endogenic modification were limited. These crustal and lithospheric conditions had a profound effect on the final morphology of many lunar basins; thus, Orientale should be used as a basin archetype only with caution.

The sites for basin-forming impacts in the lunar crust (Figure 8.1) were also chemically and petrologically heterogeneous (James, 1980; Spudis and Davis, 1986). Recognition that the lunar crust should be analyzed in terms of geochemical provinces that have distinct geological histories has given us new appreciation for

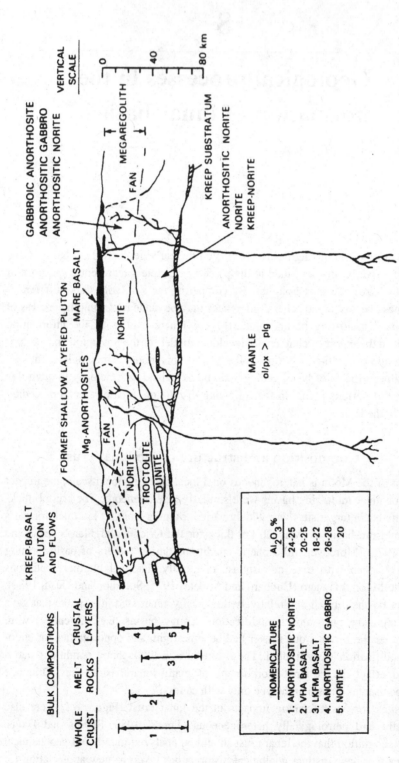

Figure 8.1 Schematic cross section of the crust of the Moon. Crust is heterogeneous vertically (crustal layers 4,5) and laterally (Mg-suite plutons); impact sites of lunar basins were similarly complex. Provenance of impact melt compositions VHA and LKFM (2,3), here interpreted as basin impact melts, shown to be middle to lower crust. After Spudis and Davis (1986).

the complexity of ejecta blankets of lunar basins and should guide attempts to relate samples collected at the Apollo sites to specific impact events (Spudis and Hawke, 1981). These regional variations may be caused by magmatic events (Longhi, 1978; James, 1980), volcanic resurfacing by KREEP and mare basalt lavas (Schultz and Spudis, 1979; Hawke and Spudis, 1980), ancient impacts that may have mobilized and removed entire crustal regions (Wilhelms, 1987), and complex combinations of any and all of the above factors. Thus, a unique chemical or petrologic signature for a given basin may not exist, with the possible exception of its sheet of impact melt. This deposit mostly is buried by mare basalt in the basins described here, but samples of either the margins of the melt sheet or ejected melt bombs are present in the sample collections (Chapters 4, 6, and 7).

Because there were older structures at the impact sites for lunar basins, structural complexities may have influenced the final morphology of a given basin significantly. Interaction with these older features can cause major departures from an idealized circular form (Chapters 3, and 7) and may accentuate or suppress regional topography in some segments of basin outer rings (sec. 4.2, 5.3). Structures tangential to basin rings, whose importance has been previously unappreciated, may result from the simultaneous effects of interaction with older topography and to flow beneath regions of thin lithosphere (sec. 4.2, 7.3). Thus, the impact targets for basins were structurally complex, chemically heterogeneous, and petrologically diverse; such factors are probably important in the development of the complexity seen in the regional geology of basins on the Moon.

8.2 Excavation

Basin-forming projectiles were abundant in the first 600 million years of the history of the Solar System (Wetherill, 1981). Although their origin is unknown, it is likely that many of the large bodies destined to collide with the planets were originally part of the disk of debris orbiting the Sun 4.5 Ga ago and that basins on the planets record the final stages of planetary accretion; occasionally, increases in the cratering rate occurred, caused by the deflection and break-up of planetesimals from the outer Solar System (e.g., Shoemaker and Shoemaker, 1990). An alternative view, originally proposed by Tera *et al.* (1974) and recently revived by Ryder (1990), suggests that the first 600 Ma of the Moon's history was largely occupied by internal geological activity and that nearly all of the large craters and basins on the Moon formed around 3.8 Ga ago, an episode known as the "cataclysm". In such a scenario, the bodies forming the lunar basins could have been derived from a co-orbiting object, a "sub-moon", which was disrupted by tidal interaction, creating a rain of objects that hit the Moon and formed the basins and cratered highlands (Ryder, 1990). For reasons dealing with the nature of the geological setting of the Apollo samples (discussed below), I prefer the former interpretation, whereby most basin-forming projectiles are derived from the outer Solar System and hit the Moon at heliocentric velocities (i.e., greater than 10–15 km/s). Such a view does not preclude significant, short-term increases in the impact flux, some of which may appear "cataclysmic".

The initial stages of basin formation, including that of early penetration of the basin projectile, are poorly understood. Results from laboratory experiments (Gault *et al.*, 1968) suggest that penetration of the projectile is accompanied by jetting of vaporized and melted target and projectile materials. Such jetted material has not been identified in the lunar samples and probably would have been ejected from the Moon at velocities much greater than the lunar escape velocity (2.5 km/s). Some fragments of late-stage jetting may be represented by some of the abundant glass spheres found in the lunar regolith at all sites, but it would be impossible to identify this material unequivocally.

The transient cavity of the basin appears when the projectile has been largely consumed and the rarefaction wave creates the cratering flow field, ejecting material from the cavity (Gault, 1974; Croft, 1981a). The geological evidence I have presented (Chapters 3–7) suggests that the transient cavity not only is a feature of much smaller diameter than that of the basin topographic rim (Figure 1.2) but also may not be expressed as an observed basin ring because of extensive modification of basin topography and burial by subsequent units. The transient cavity is *not* equivalent to the excavated cavity in depth (Dence *et al.*, 1976; Grieve *et al.*, 1977, 1981; Croft, 1981a), but *is* its equivalent in diameter (Figure 2.1). The difference in the volume of the transient and excavated cavities (Figure 2.1) consists of material that is rotated and displaced outward during the latter stages of basin excavation. This structural displacement of material implies that the inner rings of basins are composed of displaced autochthonous rocks from lower stratigraphic levels than those from which most of the ejecta are derived (Croft, 1981b).

In short, I believe that the geological evidence from study of lunar basins supports a size-invariant shape for the transient crater during impact events. This concept, called the *proportional growth* cratering model (Dence, 1977a; Grieve *et al.*, 1981; Croft, 1980, 1981a, 1985), postulates that the excavation cavity during basin formation differs only in size, not shape, from smaller cratering events. Proportional growth does *not* imply deep excavation; the maximum depths of excavation of complex craters on the Earth are about one-eighth to one-twelfth the diameter of the reconstructed transient cavity (Masaitis *et al.*, 1980; Grieve, 1987). Thus, proportional growth for basins *is* consistent with the general paucity of lower crustal and upper mantle ejecta on the Moon (Taylor, 1975, 1982; cf. Head *et al.*, 1975), long held to be incompatible with such a model (e.g., Hodges and Wilhelms, 1978). Part of the reason for a longstanding reluctance among some workers to accept the proportional growth model for basins has been a misunderstanding of the flow regimes present during crater growth (Chapter 2). The nature of the cratering flow field, with its separate transient and excavation cavities (Figure 2.1), has been fully appreciated only within the past few years.

Because basins throw out vast quantities of material and such ejection results in the redistribution of mass (creating gravity anomalies), the volume of material

Table 8.1 *Inferred cavity dimensions and total excavated volumes for five lunar multi-ring basins*

Basin	Diameter (km)[a]	D_{tc} (km)[b]	$d_{excav.}$ (km)[b]	V_{ejecta} (10^6 km^3)[c]	T_{crust} (km)[d]	V_{mantle} (10^6 km^3)	$\%V_m$[e]
Orientale	930	582 ± 77	58 ± 21	7.7 ± 2.7	90	–	–
Nectaris	860	544 ± 73	54 ± 20	6.3 ± 2.2	70	–	–
Crisium	740	488 ± 27	49 ± 13	4.5 ± 0.7	60	–	–
Serenitatis	920	572 ± 76	57 ± 21	7.3 ± 2.5	60	0-0.6	0-0.5
Imbrium	1160	685 ± 88	68 ± 25	12.5 ± 5.4	55	0.05-1.46	0.6-8.1

Notes: [a]Diameter refers to basin main topographic rim, interpreted here as structurally equivalent to rims of smaller, complex craters. Crisium main rim diameter uncertain (Chapter 5); 740 km diameter ring is current best estimate.
[b]D_{tc} – diameter of basin transient crater according to proportional growth model (Croft, 1981a, 1985; Spudis *et al.*, 1984b, 1988b, 1989). Values calculated from equation $D_{tc} = (0.47 \pm 0.05)D + (140 \pm 30)$; excavation depths – $d_{excav.} = (0.1 \pm 0.02) D_{tc}$, from Spudis and Davis (1985).
[c]Ejecta volumes (all in millions of cubic kilometers) calculated using assumed crater geometry (see text)
[d]Crustal thickness estimates at basin target sites from Bills and Ferrari (1976)
[e]$\% V_m$ = estimated fraction of mantle material in basin ejecta, according to simple, assumed excavation geometry (see text).

excavated by a basin-forming impact might be a test for models of basin formation. I have estimated the total volume of material ejected from the excavation cavity for the five basins described in this book (Table 8.1). My model consists of a spherical Moon excavated by a hemispherical cavity to a depth equal to one-tenth its diameter (Croft, 1981a). Although the true shape of the excavation cavity is defined by flow streamlines of spiral form (Maxwell, 1977; Croft, 1980, 1981a), the approximation of this geometry by a hemispherical cap does not significantly affect the volume estimates (S.K. Croft, personal communication, 1981).

The range of estimates of ejecta volumes (Table 8.1) results from the uncertainty of the true size of the original crater; in some instances, this variation is quite large (e.g., Imbrium). I estimated the total amount of material excavated from the lunar mantle during basin formation, on the basis of approximations of crustal thickness from Bills and Ferrari (1976). Two of the basins may have been large enough to produce sub-crustal (mantle) ejecta: Serenitatis and Imbrium (Table 8.1). Both the Serenitatis and Imbrium basins have large quantities of materials from the lower crust in their ejecta (Chapters 6 and 7) and mantle fragments could be contained within deposits that are partly buried by mare basalt flooding, within portions of their ejecta that were not sampled by a mission, or those that are not covered by remote-sensing data.

These estimates of ejecta volumes for lunar basins (Table 8.1) are comparable to

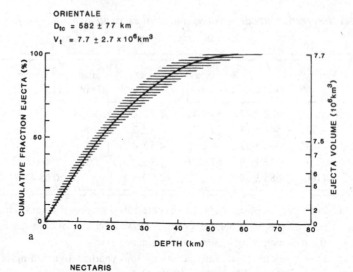

ORIENTALE
$D_{tc} = 582 \pm 77$ km
$V_t = 7.7 \pm 2.7 \times 10^6$ km^3

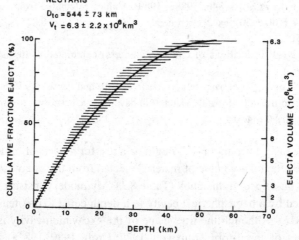

NECTARIS
$D_{tc} = 544 \pm 73$ km
$V_t = 6.3 \pm 2.2 \times 10^6$ km^3

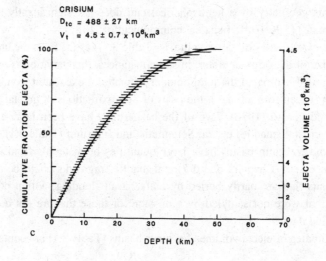

CRISIUM
$D_{tc} = 488 \pm 27$ km
$V_t = 4.5 \pm 0.7 \times 10^6$ km^3

those derived from other methods. Moore *et al.* (1974) used photogeology to estimate an excavated volume for Orientale of 4.5×10^6 km^3. A separate analysis by Scott (1974) analyzed the apparent mass deficit of the Orientale basin from orbital gravity data; he estimated that 5.3×10^6 km^3 of material was excavated. More recently, Bratt *et al.* (1985) modeled the amounts of structural uplift of the crust–mantle boundary to estimate excavated volumes; they found a total excavated volume of 7×10^6 km^3 for the Orientale basin. These values may be compared with Table 8.1; I estimate that the Orientale basin excavated $(7.7 \pm 2.7) \times 10^6$ km^3

Figure 8.2 Cumulative ejecta volumes of five lunar basins, estimated from a simple hemispherical cavity geometry. Shaded areas indicate uncertainty of reconstructed diameters of transient cavities. A consequence of the shape of excavation cavity is that virtually all basin ejecta are of crustal provenance. (a) Orientale; (b) Nectaris; (c) Crisium; (d) Serenitatis; (e) Imbrium.

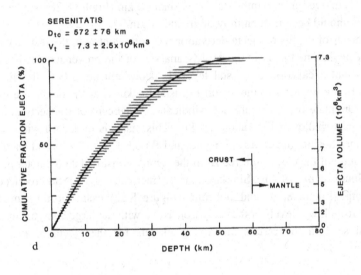

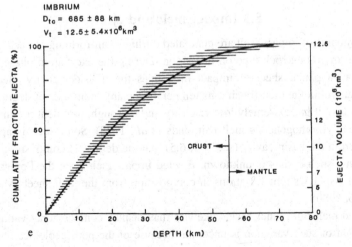

of crustal material. Thus, the proportional growth model applied to the basins studied in this book predicts total estimated volumes that agree with observations for the Orientale basin.

Although basin-forming impacts excavate deeply from the Moon, they are not large enough to excavate much material below the lunar crust. This shallow excavation may explain one of the most puzzling aspects of the Apollo sample collection, the apparent absence of material from the lunar mantle (Taylor, 1975, 1982). To address the provenance of basin ejecta with depth, I used a simple hemispherical geometry for the crater of excavation to generate the curves presented in Figure 8.2; these plots show the cumulative amounts of ejecta that come from given depths within the lunar crust. Although lunar basins excavated to as deep as 93 km in the case of Imbrium, the vast bulk of ejecta is derived from much shallower stratigraphic levels. For the maximum size of the Imbrium basin, 90 percent of the ejecta volume is derived from depths shallower than 60 km (Figure 8.2e), suggesting that its ejecta should be almost entirely of crustal origin.

Estimates of depths of ejecta derivation are consistent with the results of mixing model calculations for basin ejecta. My analysis of the provenance of basin ejecta predicts that excavation of crustal levels is dependent upon both the size of the basin-forming impact and the average crustal thickness at the target site. Analysis for the Orientale basin (Figure 8.2a) indicates 90 percent of its ejecta are derived from depths shallower than about 35 km. This result is in accord with the dominantly anorthositic (upper crustal; Ryder and Wood, 1977; Spudis and Davis, 1986) composition of the ejecta inferred from the remote-sensing data (Chapter 3; Spudis, 1982; Spudis *et al.*, 1984b). Similarly, larger fractions of the ejecta produced by the Serenitatis (Figure 8.2d) and Imbrium (Figure 8.2e) basin-forming impacts are derived from lower levels of the crust, consistent with the large component of mafic material seen in the results of mixing models for the ejecta of these basins (Chapters 6 and 7).

8.3 Impact melt and ejecta

Large amounts of impact melt are generated during basin-forming impacts and constitute an important rock type on the Moon. During the excavation phase of basin formation, a planar sheet of impact melt lines the transient cavity. Because of superheat within the melt (raising its temperature many thousands of degrees above the liquidus), it has extremely low viscosity and thorough, turbulent mixing creates a chemically homogeneous melt (Simonds *et al.*, 1978). Some melt is apparently ejected from the basin cavity (sec. 4.4, 6.5), but the degree of compositional variation within such bodies is unknown. Ejected impact melt from the Popigay crater on Earth displays slight variations in composition from the melt sheet (Masaitis *et al.*, 1976, 1980).

Ejected melt from lunar basins likewise may display compositional variation, but the amount of such variation is uncertain because of the poor geological control of lunar samples. If the Apollo 17 melt "sheet" is ejected melt from the Serenitatis basin (sec. 6.5), then such ejected melt *is* homogeneous and may be derived from the margins of the main melt sheet. I believe it more likely that melt entrained within clastic ejecta in the continuous deposits of basins can display varied compositions, a phenomenon I infer from the variation seen in the Apollo 16 VHA impact melts (Figure 4.11; Spudis, 1984; McKinley *et al.*, 1984), which are interpreted here as impact melts of the Nectaris basin (sec. 4.4). Basin melt may be ejected in a relatively continuous stream of slightly varied composition (Ryder and Wood, 1977) and this stream may not have a composition identical to the planar sheet of melt that remains on the basin floor (e.g., the Maunder Formation of the Orientale basin; Chapter 3).

Impact melts from the Apollo landing sites have varied composition and multi-

Figure 8.3 Concentration of Ti and Sc (two refractory elements) in basaltic impact melts from three highland landing sites on the Moon. Different groups are distinguished at each site: melt groups 1–3 at Apollo 16 are defined by McKinley *et al.* (1984), groups "Poik" and "Aph" are the two melt categories at Apollo 17 (e.g., Spudis and Ryder, 1981), groups A–E are found at the Apollo 15 site (Ryder and Spudis, 1987). The field of the melt sheet from the terrestrial Manicouagan crater is shown for comparison. Although showing considerable overlap, melts from a given site tend to have compositions that are (1) distinct from the local upper crust, implying exotic derivation (here interpreted as being from basins), and (2) distinct from each other, implying creation in different basin-forming impacts.

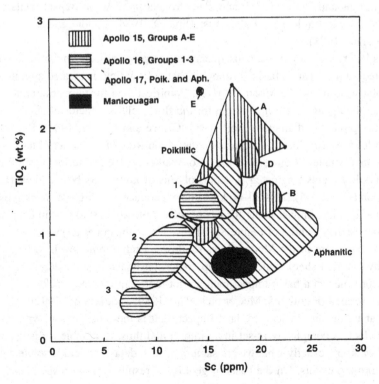

ple melt groups occur at each highland site, but there is a tendency for the compositions of impact melt rocks to cluster at a given site (Figure 8.3). I believe that this clustering reflects the dominance of impact melt from a single basin at the Apollo 16 (Nectaris), Apollo 17 (Serenitatis), and Apollo 15 (Imbrium) sites. The compositional mapping of basin-derived components is blurred, both by the presence at each site of impact melt of purely local provenance (e.g., Group 3 at Apollo 16) and by the possible presence of basin melts at distant sites (e.g., melt from Imbrium basin at Apollo 17). If my supposition is correct, then the extreme compositional homogeneity observed in the melt sheets of terrestrial impact craters (e.g., Grieve and Floran, 1978) is *not* a guide to the interpretation of lunar impact melts. In other words, diffuse compositional clusters do not *necessarily* indicate multiple impacts on the Moon (e.g., Ryder, 1981a), although such a record is not excluded.

The composition of impact melt from basins is an important source of information for both the structure of the lunar crust and melt petrogenesis. The impact melts of lunar basins have a basaltic composition (i.e., VHA or LKFM basalt; secs. 4.4, 6.5, 7.6), one that is quite distinct from that of the average lunar surface (Davis and Spudis, 1985, 1987; Spudis and Davis, 1986). Clastic debris contained in these melts does not represent the precursor materials (protolith) from which the impact melt formed; there are missing components rich in transition metals (e.g., Ti, Sc) and KREEP (Spudis *et al.*, 1991). These relations suggest that (1) the lower crust of the Moon is more mafic in composition than the upper crust (Ryder and Wood, 1977; Spudis and Davis, 1986), and (2) petrogenesis of basin impact melts occurs at middle to lower levels of the crust (Spudis, 1984; Spudis and Davis, 1986; Spudis *et al.*, 1991).

This latter implication is a consequence of the proportional growth model and is predicted by it. During a basin-forming impact, a flow field is created around a central point well below the Moon's surface. Empirical data for terrestrial craters indicate that the apparent depth of origin for the flow field is a factor of about 1 to 1.5 times the projectile diameter (Croft, 1980; Grieve and Garvin, 1984). Because the projectiles for basin-forming impacts on the Moon were on the order of tens of kilometers in diameter (Figure 8.4; see also Wetherill, 1981), during basin-forming impacts, flow fields originated at depths of tens of kilometers below the surface of the Moon (Figure 8.5). This 20–60 km zone experiences the highest shock pressures (Chapter 2); thus, impact melts for basins were generated 20 to 60 km deep within the Moon (zones 2 and 3 of Figure 8.1). However, as the crater grows outward laterally, the curvature of the Moon allows shallower stratigraphic levels to be encountered by the melt sheet, which picks up cooler, unshocked clastic debris. Thus, the clasts contained in a basin impact melt do not represent the deep rocks from which the melt formed (Figure 8.5; McCormick *et al.*, 1989; Spudis *et al.*, 1991).

Lunar basins are difficult to date. Impact melt is the only product whose radiometric clock is completely re-set in an impact and thus, is suitable for isotopic dating. If we know exactly what we are sampling, such data can yield absolute ages for basin-forming events. On the basis of geological results from the Apollo sites near the Nectaris (Chapter 4), Serenitatis (Chapter 6), and Imbrium (Chapter 7) basins, we can estimate the ages of these features. I interpret the VHA impact melts from the Apollo 16 landing site as melt from the Nectaris basin; dating of these rocks suggest, that they formed 3.92 Ga ago. The Apollo 17 poikilitic melt sheet probably represents melt from the Serenitatis basin; these rocks formed 3.87 Ga ago.

For the Imbrium basin, poor plateaux for black-and-white rocks 15445 and 15455 prevent clear ^{40}Ar/^{39}Ar age determination for that impact on the basis of its impact melt. Other impact melts from the Apollo 15 site have good age determinations that range from 3.83 to 3.88 Ga (Dalrymple and Ryder, 1991), but the relation of these melt rocks to the Imbrium basin is unknown. The volcanic KREEP basalts from the Apollo 15 site are well dated at 3.85 ± 0.05 Ga (Carlson and Lugmair,

Figure 8.4 Sizes of the basin-forming projectiles. Energies of basin formation are estimated to be between 10^{32} and 10^{33} ergs (Baldwin, 1963; Gault *et al.*, 1975). At r.m.s. impact velocities for the Moon (20 km/s; Shoemaker, 1977), chondritic, spherical projectiles are on the order of 30 to 50 km in diameter.

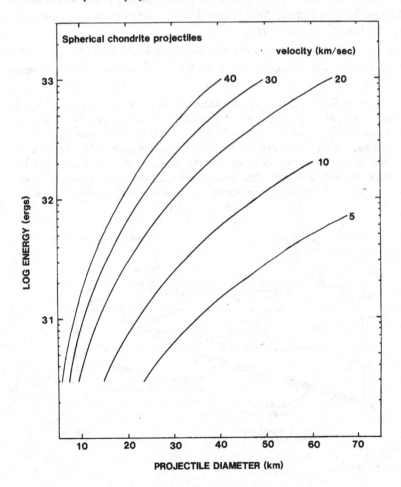

1979). Because these KREEP basalts were extruded very soon after the Imbrium impact (Spudis and Hawke, 1986; Ryder, 1987), this age must be a lower limit for the age of the Imbrium basin. I reject the younger age of 3.75 Ga for the Imbrium basin proposed by Deutsch and Stöffler (1987) and Stadermann *et al.* (1991) on the basis of geological and petrological considerations (secs. 7.6 and 7.7).

These absolute ages for lunar basins are working models. Given the errors associated with methods of radiometric dating, it is possible that all of the basins discussed in this book formed within a few years of each other, between 3.8 and 3.9 Ga ago. Such a clustering of basin ages, and the associated compression of the Nectarian-age cratering record of the Moon into a narrow, 100 Ma interval, would certainly deserve the appellation "cataclysm" (Ryder, 1990). Radiometric dates of clasts of impact melt in meteorites from the Moon are a key piece of evidence bearing on the reality of the cataclysm. Because these rocks probably come from areas distant from the Apollo sites (Warren, 1991), a prevalence of 3.8–3.9 Ga ages in the meteorites would indicate that significant cratering occurred over the whole Moon at that time. This result would negate one of the principal arguments against the cataclysm, the idea that the Apollo ages date only one or a few large events (e.g., Wetherill, 1981; Wilhelms, 1987).

The basin age assignments presented above are in contrast to those of Baldwin (1987a, 1987b), who contends that both the Nectaris and Serenitatis basins are

Figure 8.5 Schematic drawing showing the stratigraphic provenance of impact melts of lunar basins and their incorporated clastic debris. Assuming depths of penetration of about one projectile diameter (Croft, 1980), most basin melt originates in the lower crust of the Moon. Ejected melt sampled by Apollo missions leaves the excavation cavity such that entrained clasts are from middle to upper levels of the crust. Thus, impact melts of lunar basins come from rock types different than those from which their contained clastic debris are derived (Spudis *et. al.*, 1991).

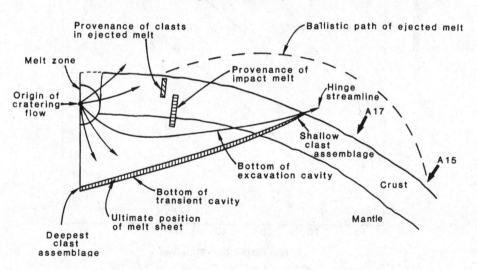

much older (> 4 Ga old) than my estimates. Baldwin (1987a,b) used the rim topography of basins and crater counts to reach these conclusions. As discussed in Chapters 4 and 6, topographic rejuvenation may have occurred at both of these basins, along with anomalous subsidence in some areas. If this happened, it would introduce significant error into Baldwin's calculations. Moreover, many of the craters in the highlands surrounding the Serenitatis basin are secondary craters from the Imbrium impact (sec. 6.2). I interpret the Nectaris, Serenitatis, and Imbrium basins to have formed 3.92, 3.87, and 3.85 Ga ago, respectively.

Calculations of the cratering process (Schultz, 1978; Schultz *et al.*, 1981) suggest that ejecta within the transient cavity remain there longer with increasing crater size. Therefore, ejecta from large basins probably remain within the transient cavity for very long times in comparison with smaller impact events, probably on the order of tens of minutes, as opposed to tens of seconds. Such longevity in residence within the cavity would be expected to increase the efficiency of comminution (Gault *et al.*, 1963; Schultz and Mendell, 1978) such that the median grain sizes for basin ejecta may be much smaller (probably on the order of centimeters to tens of centimeters) than those of ejecta from large lunar craters (Schultz, 1978). Such a fine grain-size greatly complicates the task of identifying primary ejecta from lunar basins in the Apollo samples.

Ejection angles for material excavated during basin-forming impacts on the Moon are unknown. If jetting occurs during basin impacts, the initial angle of ejection upon projectile penetration is probably quite low, on the order of a few degrees above the horizontal (Melosh, 1989), but at very high velocity. Ejection during the main phase of cavity excavation was probably at higher angles; Oberbeck (1975) suggests that 15 degrees is a good estimate, on the basis of laboratory experiments. As excavation nears completion, the rebounding floor of the basin would tend to lower this ejection angle. However, these low angles of ejection are not those of the ejecta as they land back on the Moon. Because of the abrupt curvature of the Moon at basin scales, distal margins of the ejecta curtain would intersect the lunar surface at increasingly steeper angles; toward the edge of the continuous deposits, the ejecta curtain could be nearly vertical to the surface (Schultz *et al.*, 1986). Thus, radial momentum and subsequent surface flow of basin deposits probably are not caused by extremely low ejection angles (Chao *et al.*, 1975; Wilhelms, 1987). I interpret the flow textures in basin deposits as resulting from surface movement radially outward after ballistic emplacement (Chapter 7). Additionally, hummocky materials that occur near the rims of basins (e.g., the Apenninus material of the Imbrium basin; see Chapter 7), cited by Wilhelms (1987) as evidence for low-angle ejection, may instead be related to slumping following collapse and ring formation (Spudis and Head, 1977).

Continuous basin deposits consist of primary ejecta mixed with materials derived locally by secondary cratering during emplacement. I believe that basin deposits probably contain a higher fraction of primary ejecta than predicted by the model of Oberbeck (1975), because in my opinion, primary ejecta from Imbrium were sampled at the Apollo 14 site (sec. 7.5) and ejecta from Nectaris were sampled at the Apollo 16 site (sec. 4.4). Moreover, the fine size of clastic ejecta from basins and the finite thickness of their ejecta curtains (Schultz, 1978; Schultz and Gault, 1985) suggest an extended time of deposition for basin ejecta. Such extended deposition would promote the dilution of local materials (Morgan *et al.*, 1974). Local mixing becomes more efficient in the discontinuous deposits of lunar basins, as shown by the strong dependence of regional compositions in the central highlands on local geochemical provinces, an area dominated by the discontinuous facies of the Imbrium basin (secs. 4.1, 7.7).

Light plains, such as the Cayley Formation (sec. 7.2), are made up at least partly of primary ejecta from basins, as this material may mantle older flows of mare basalt (Schultz and Spudis, 1979, 1983). The ratio of local to primary ejecta in the Cayley plains is not known, but mottling of the spectral signature of light plains in the Schiller–Schickard region could indicate admixture of local, pre-Orientale mare basalt with primary ejecta from the Orientale basin (Hawke *et al.*, 1991). Thus, discontinuous deposits are dominated by locally derived materials where large secondary craters are evident, in the distal deposits of basins.

Impact melt from basins may be partly ejected and partly included within the continuous deposits. Thus, continuous basin deposits are a two-phase mixture containing mostly clastic debris and subordinate impact melt. Such a relation does *not* imply that the ejecta blanket as a whole was hot; ejected melt is localized spatially, surrounded by much cooler clastic material, and thus, thermal metamorphic effects are extremely limited. Some of this melt may segregate from the debris to form flow lobes or melt ponds (Chapter 3). Other melt masses may be finely dispersed or energetically mixed with clastic debris upon deposition to form melt-poor, clast-rich polymict breccias (Spudis, 1981).

The melt sheet of the basin largely remains in place during the excavation and modification stages of basin formation. This melt mantles the uplifted floor in the basin center, producing the fissured and fractured melt sheet seen in the interiors of fresh, unflooded basins (e.g. Orientale; Figure 3.4; Chapter 3). Tongues may segregate from the main melt sheet, forming ponds or pools (Chapters 6, 7); this melt would be of identical composition to the main melt sheet because it was generated in the same environment (cf. ejected melt masses represented by the VHA impact melts of Apollo 16; sec. 4.4). Thus, both the Apollo 15 black-and-white rocks (sec. 7.6) and the Apollo 17 poikilitic melts (sec. 6.5) are representative of the melt sheets produced by the Imbrium and Serenitatis basin-forming impacts, respectively. I summarize the relations among units of basin ejecta sampled by the Apollo and Luna missions in Figure 8.6.

The morphology of basin deposits is controlled by the energy of its environment of deposition. Near the rim of the basin, late-stage excavation is low-energy and units display hummocky, dune-like morphology (e.g., Descartes of Nectaris, sec. 4.2; Apenninus material of Imbrium, sec. 7.2). Away from the basin rim, the higher-energy environment producing the continuous deposits forms undulating, knobby, or smooth deposits, such as the Fra Mauro and Alpes Formations of the

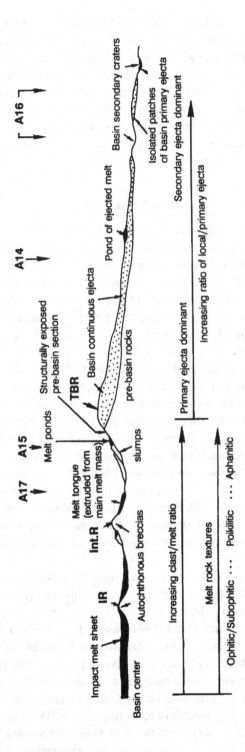

Figure 8.6 Inferred geological relations for ejecta deposits of lunar basins. Inner ring, intermediate ring(s), and topographic basin rim are shown (see Figure 1.2). Numbers at top give relative positions of Apollo landing sites to deposits of basins; Apollo 16 site shown relative to Imbrium. The position of Apollo 16 relative to Nectaris is the same as that of Apollo 14 to the Imbrium basin.

Imbrium basin (section 7.2). Isolated patches of primary ejecta from the basin that occur within the discontinuous deposits may also have this appearance (e.g., the Sculptured Hills material on the eastern rim of Serenitatis, partly related to the Imbrium basin; sec. 6.2). The depositional environment with the highest energies is associated with the discontinuous facies of basins and consists of large secondary craters, crater chains and clusters, and smooth plains. In this zone of basin deposits, the volume of locally derived ejecta from secondary craters exceeds the amount of primary ejecta (Oberbeck, 1975) and energetic mixing of the two components produces a complex, polymict deposit.

8.4 Ring formation

The excavation of such a large volume of material (Table 8.1) from the lunar crust during a basin-forming impact must have had catastrophic consequences. One of the most interesting results of basin formation is the development of multiple, concentric rings. The origin of basin rings remains the greatest missing piece of the basin puzzle. However, on the basis of results presented here, some geological constraints may be placed on the origin of rings. Part of this process entails a consideration alluded to in Chapter 2, i.e., the spacing of rings in lunar basins.

Hartmann and Kuiper (1962) were the first to note that distance between adjacent rings of lunar basins increased outward from the basin center; they claimed that this increase was by a constant factor, which they believed to be about 1.4 or 2.0, depending on which basin and which rings were considered. Picking up on this observation, Fielder (1963) maintained that the rings of Orientale, Humorum, Nectaris, and Imbrium were all spaced at a constant factor, and this factor, specifically, is $\sqrt{2}$; he also noted that not all ring positions had to be occupied by a ring for the $\sqrt{2}D$ spacing "rule" to hold (see Table 4 of Fielder, 1963). Later, Hartmann and Wood (1971) extended this "$\sqrt{2}$ spacing rule" to include *all* multi-ring basins on the Moon, identifying 27 basins with over 69 rings, most of which are spaced at this factor (many of the features designated as basins in this paper are protobasins (Pike, 1983), and thus not included in my inventory of lunar basins in Tables 2.1 and 2.2).

The physical significance of a spacing increment for basin rings at $\sqrt{2}$ was, and is, unknown. Authors can only agree that it must be important (for example, a factor of $\sqrt{2}$ repeatedly occurs in nature in a variety of periodic and wave functions). Fielder (1963), Hartmann and Wood (1971), and Schultz (1976a) suggested that fractures induced by bending steel plates, described by Lance and Onat (1962), are mechanical analogs to basin rings; however, Pike and Spudis (1987) examined the Lance and Onat experiments and demonstrated that these results do not apply to basin rings. Subsequent workers have largely tended to ignore the "square root of two" rule in developing their models of basin ring genesis. Those that did incorporate it found it difficult to accommodate; Van Dorn (1968), developing a ring model for Orientale based on Baldwin's "tsunami" mechanism (Baldwin, 1974,

1981), found that he had to invoke *ad hoc* a layered structure for the Orientale target to explain the $\sqrt{2}\,D$ spacing of Orientale basin rings.

The crucial questions in regard to the spacing of basin rings are (1) Are basin rings spaced at the factor $\sqrt{2}\,D$? and (2) If they are, what are the implications of this relation for ring genesis? The most comprehensive study of these two questions to date is by Pike and Spudis (1987). In this work, we critically tested the $\sqrt{2}\,D$ spacing rule for all multi-ring basins on the Moon, Mercury, and Mars; our results for the Moon are shown in Figure 8.7. In short, we found that lunar basin rings (and the rings of basins on the other planets) *are* in fact spaced at this interval of $\sqrt{2}$; this spacing rule was shown to be true at the 99 percent confidence level (Pike and Spudis, 1987). Because the rule holds for three different planets, with differing gravities and target conditions at the time of impact, the location of basin rings is somehow related to physical conditions associated with the impact itself. We were careful not to assert that basin rings *form* at the time of impact, only that the sites of future ring *location* were established at this time (Pike and Spudis, 1987).

Given that this $\sqrt{2}$ spacing rule holds, any ring-forming model must be consistent with such spacing. As mentioned above, a mechanism to produce basin rings that is compatible with all available constraints has not yet emerged. The three categories of ring-forming mechanisms (Table 1.2) all have arguments for and against them. In the following paragraphs, I will not attempt to solve the ring problem, but rather, will try to elucidate some of the elements that a comprehensive ring-forming model must incorporate.

The topographic rim of basins is equivalent to the rims of smaller, complex craters (e.g., Copernicus). Such equivalence has long been assumed by most workers, including those that do not adhere to proportional growth models of basin formation (e.g., Wilhelms, 1987, p. 78). Wilhelms (1987) and I differ in that he believes that the main rim of craters marks the boundary of impact excavation. This supposition equates an ephemeral feature, the transient cavity (produced during the excavation phase of basin formation) with a permanent feature, the crater rim (produced during the modification phase). Study of both experimental (e.g., Stöffler *et al.*, 1975) and terrestrial impact craters (e.g., Grieve, 1987) demonstrate that these two phases are in fact distinct, even though they may overlap in time.

Immediately following the basin excavation phase, the region surrounding the excavation cavity is deformed and moves upward in response to the crustal unloading produced by excavation. This deformation and crustal adjustment produces at least some basin rings. I believe that the geological evidence indicates that the topographic rim of lunar basins is formed largely by structural uplift and subsequent collapse, specifically, by the formation of a megaterrace (Mackin, 1969; Gault, 1974; Head, 1974a, 1977a; Dence, 1976). The fault origin of the Orientale Cordillera ring is evident in the deformation and offset of older structures in that region (Chapter 3). The scarp-like appearances of both the Cordillera range of Orientale (sec. 3.3) and the Altai ring of Nectaris (sec. 4.2) are strongly suggestive of a structural origin. A structural exposure of pre-Imbrian rocks within the

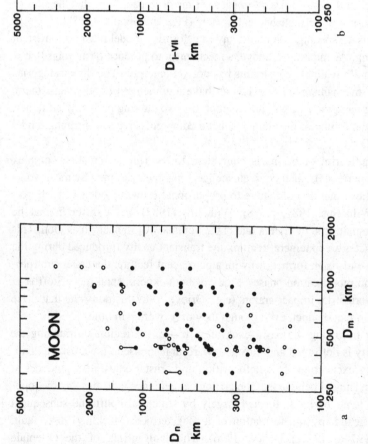

Figure 8.7 Systematic spacing of rings in multi-ring basins on the Moon (data in Table 2.1). From Pike and Spudis (1987). (a) Raw data for unranked rings. D_m is the diameter of the main topographic rim (chosen by photogeological analysis) of the basin. D_n is the diameter of any other observed ring. Open circles are less certain ring diameters; connected circles are split rings (see Pike and Spudis, 1987). Several sub-parallel trends are evident. (b) Ranked ring data showing 95% confidence intervals for calculated linear best-fit lines. Overall trend is at least six sub-parallel lines, one for each ring rank (basin rim = IV). All lines slope at near-unity; basin rings on the Moon are spaced at the $\sqrt{2} D$ interval. For details, see Pike and Spudis (1987).

Table 8.2 *Proposed synthesis for basin formation*

Constraints on basin excavation

(1) Basin ejecta largely of crustal origin; little evidence for mantle material.

(2) For Orientale, V_{ejecta}= 4 to 8×10^6 km^3 (gravity, photogeology).

(3) Pre-basin structures preserved *within* main topographic rim (Orientale, Imbrium).

(4) Basin impact melts LKFM or VHA; more mafic than average surface composition.

Constraints on basin ring formation

(1) Main topographic rim of basin is structural equivalent to rim of complex craters

(2) Outer basin rings exist; age relation to host basin uncertain

(3) Inner *and* outer basin rings spaced at $\sqrt{2}\,D$ intervals

(4) Evidence for deep-seated structures associated with rings

(5) Outcrops of pure anorthosite on Moon largely confined to inner rings of basins

Basin-forming model

(1) Proportional growth valid; $D_{tc} = (0.47 \pm 0.05)D + (140 \pm 30)$, in kilometers.

(2) Maximum depth of excavation, $d_{excav} = (0.10 \pm 0.02)D_{tc}$

(3) Volume of ejecta for Orientale 7.7×10^6 km^3

(4) Basin impact melts are VHA and LKFM compositions, generated at middle to lower crustal levels (30–60 km depth)

(5) Main rim formation by collapse around smaller excavation crater (megaterrace)

(6) Basin ring *positions* (not necessarily topography) determined at time of impact (resonance effects?)

(7) Inner rings of basins formed by structural uplift of deep crustal levels

(8) Lithospheric fracturing and failure associated with major impacts (outer rings?)

Source: after Pike and Spudis (1987)

Apennine front (sec. 7.6) supports a megaterrace origin for this part of the rim of the Imbrium basin.

The intermediate ring (Figure 1.2) of basins may also be of structural origin, because older topography is often preserved inside of the main ring (secs. 3.3, 7.3) and sometimes inside the intermediate ring (sec. 3.3). Such preservation would not be possible if the tsunami (Van Dorn, 1968; Baldwin, 1974, 1981) or nested-crater models (Wilhelms *et al.*, 1977; Hodges and Wilhelms, 1978) were responsible for ring formation. Inner rings of basins (whose diameter is smaller than the inferred diameter of the transient cavity) may form either by forceful oscillatory uplift (Murray, 1980; Grieve *et al.*, 1981) or by structural uplift, followed by collapse (Head, 1977a; Melosh and McKinnon, 1978; Schultz *et al.*, 1981); either mechanism is compatible with the observed geological relations. The documentation of minor oscillatory movement in well characterized craters on Earth (Grieve *et al.*, 1981) suggests that this mode of origin for inner rings may be likely for lunar basins.

The origin of the outer rings of basins, exterior to the main rim (Figure 1.2), is

For legend see page 186.

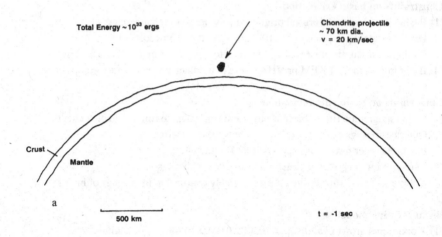

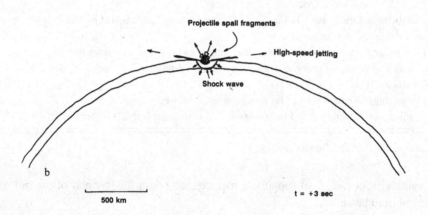

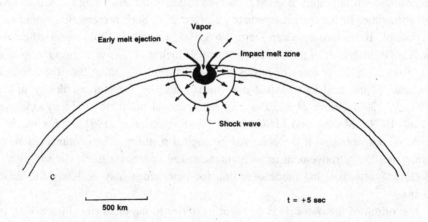

For legend see page 186.

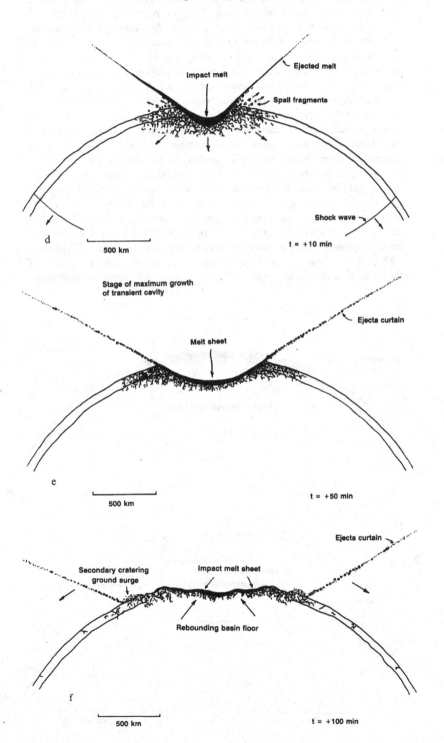

Figure 8.8 Sequence showing stages in the formation of the lunar Imbrium basin (to scale). Time relative to the impact (in seconds and minutes) shown in lower right of each panel. (a) Just prior to impact. Assuming a chondritic body hitting the Moon at lunar r.m.s. velocity (about 20 km/s), the projectile was probably on the order of 70 km in diameter. Total energy of formation for Imbrium was on the order of 10^{33} ergs (Baldwin, 1963). (b) Just after contact of the projectile with the Moon. The shock wave propagates both into the Moon and the projectile, causing spall from the back side of the projectile. Very high speed jetting of melt and vapor at the projectile–Moon contact also occurs at this stage. (c) After the projectile has been largely consumed, a cratering flow field is created, while the shock wave continues its propagation into the body of the Moon. Note that the zone impact melting is largely confined to lower crustal and upper mantle materials, resulting in the formation of low-K Fra Mauro impact melt (e.g., Spudis, 1984). (d) Excavation of the basin continues, largely resulting from lateral growth of the cavity. Arrival of the cavity wall is preceded by surface spall, most of which leaves the Moon at greater than escape velocities (2.5 km/s). (e) The stage of maximum growth of the transient cavity, about an hour after initial contact of the projectile. Zone of intense fracturing surrounds transient crater and may be "acoustically fluidized" (Melosh, 1979). (f) In response to the large amount of crustal unloading, the bottom of the cavity rebounds rapidly, while a sheet of impact melt covers the basin floor.

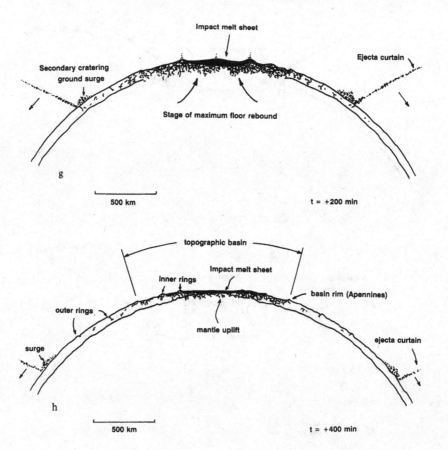

still enigmatic. These rings follow the same $\sqrt{2}$ spacing rule as interior rings (Figure 8.7), are usually less prominent topographically, and frequently display such structural features as vents for the extrusion of mare basalts (Chapter 7). These relations suggest that the outer rings are scarps of structural origin, formed along zones of weakness established during the impact. Their age relative to the basin event is unknown; perhaps they form by lithospheric fracturing on a long-term basis, several million years after the basin impact (e.g., Hartmann and Wood, 1971; Melosh and McKinnon, 1978).

The scenario for the formation of basin rings outlined above resembles none of the classic models of ring formation (Table 1.2) in pure form, but rather, embodies certain elements from all of them. The main rim of the basin is dominantly a feature formed by structural collapse (megaterrace), inner rings are created by uplift followed by collapse (oscillating peaks), and outer rings reflect collapse of the lithosphere or basin target after the impact event (target strength). In my opinion, the most realistic model for ring development advocated to date is that of Grieve *et al.* (1981); this model incorporates elements of a variety of different mechanisms into a consistent and coherent scenario of the development of basin rings.

The scenario I have developed for the formation of multi-ring basins on the Moon is summarized in Table 8.2 (Pike and Spudis, 1987) and the sequence of events in the formation the Imbrium basin is illustrated in Figure 8.8. This summary should be considered a "progress report", as not all of the problems posed by multi-ring basins have been solved completely. However, lest the reader think that this has been an exercise in futility, it should be noted that several of the major questions raised about basin formation during post-Apollo study of the Moon have been resolved: Ejecta from the deep crust or upper mantle are rare on the Moon because little, if any, was excavated by basins. The main rim of basins *is* equivalent structurally to the rims of smaller, complex craters. Transient cavities are smaller than the presently observed rims of the basins and excavation of material is con-

---

Legend 8.8 continued.

Material in the ejecta curtain begins to be deposited on the lunar surface, beginning with the area nearest the crater rim. The ejecta curtain is followed by a large zone of material that has been kicked up by the impact of the curtain (the ground surge of secondary crater ejecta; see Oberbeck, 1975). (g) The basin floor shows a maximum amount of rebound, possibly over-steepened with respect to the original curvature of the Moon. This configuration is unstable and begins to collapse. Deposition of material from the ejecta curtain continues at greater ballistic ranges, followed by the surge of secondary crater ejecta. (h) Collapse is largely complete, with major interior and exterior rings having formed. Sheet of impact melt is still nearly completely molten and lines the basin floor. The last portions of the ejecta curtain are being deposited at extreme ballistic ranges, but only in a discontinuous, fragmentary manner. Continued structural adjustment of portions of ring system may be expected for many years after this stage (see Figure 8.9).

fined to upper levels of the crust. Impact melts produced during basin-forming impacts are more mafic than the composition of the average lunar surface because they are generated at middle to lower levels in the crust. Basin rings are spaced at a constant factor of $\sqrt{2}$ and many basin rings are of dominantly structural origin. Detailed exploration of all the ramifications of current understanding of basin formation is still ongoing and lunar research is an active and productive field of inquiry.

8.5 Long-term modification of basin topography

For the older lunar basins, modification of basin topography continued long after the impact. This modification is largely a result of viscous relaxation of basin topography over geological time (Solomon *et al.*, 1982). Because the Moon was very hot early in its history, the topographic relief of the older basins deformed in a plastic manner. This deformation reduced basin relief and accentuated platform massifs as the surrounding crust subsided. Such relaxation was apparently much more efficient for basins on the near side than those on the far side of the Moon; Solomon *et al.* (1982) found that the Tranquillitatis basin shows substantial evidence for relaxation of topography whereas the relief of the South Pole–Aitken basin is nearly pristine. The concept that the strength and/or thickness of the lunar lithosphere was laterally variable is also supported by the "transitional" nature of the Serenitatis and Imbrium basins, portions of which resemble old, Crisium-style rim topography, while other rim sectors resemble the rings of Orientale (Chapter 3).

In addition to viscous relaxation, internal modification by resurgence, as described for floor-fractured craters by Pike (1971) and Schultz (1976b), also may have affected lunar basins, at least to a minor extent. At Nectaris, rejuvenation of the Altai scarp may be responsible for its topographic prominence (sec. 4.2). The ring troughs of the Crisium basin and mare flooding of zones between platforms suggest long-term development of that basin by structural deformation and igneous activity (sec. 5.6). In addition, both seismic disturbances (e.g., Schultz and Gault, 1975) and deposition of ejecta from subsequent basin impacts (e.g., Serenitatis, Imbrium) may have reactivated structural weaknesses in the Crisium region and enhanced the modification of topography. Such processes may also have been important to a lesser degree at the transitional Serenitatis and Imbrium basins.

The basins discussed in this book passed through distinct courses of modification that reflect their different ages and locations on the Moon (Figure 8.9). The older basins Crisium and Nectaris, as well as portions of Serenitatis and Imbrium, formed in a relatively thin lithosphere. In this environment, rigid crustal blocks were deformed over a plastic substrate, producing the characteristic platform topography associated with some of the basin massifs (Figure 5.12; Chapters 4–7). This deformation resulted from structures radial and concentric to the basin and such fractures served as vents for the eruption of lavas in some cases (e.g., interplatform volcanic flooding in the highlands surrounding Crisium, sec. 5.2). The younger

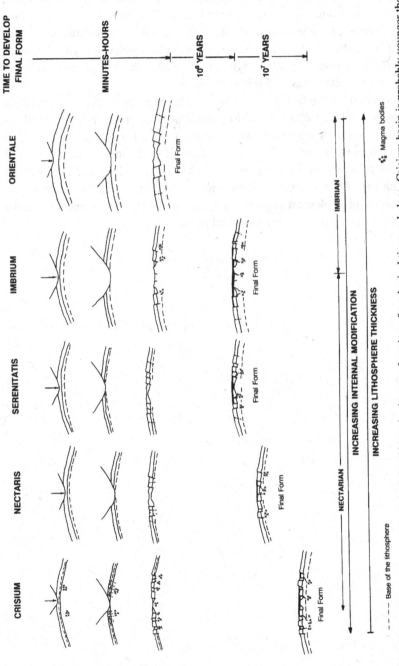

Figure 8.9 Differences in the modification style of lunar basins as a function of geological time and place. Crisium basin is probably younger than Nectaris (sec. 5.2), but its impact site had a relatively thin lithosphere; thus, its style of long-term, internal modification resembles older basins. The change in the style of modification of basins with time is towards rapid completion of final morphology for the younger impacts; this trend is a consequence of a gradually thickening lithosphere, resulting from global cooling of the Moon.

Orientale basin, and parts of Serenitatis and Imbrium, formed in a thick, rigid lithosphere and deformation caused by sub-lithospheric flow was not significant. These basins closely resemble craters and hummocky to blocky massifs make up the rings of these basins, similar to the morphological elements of the smaller pro-tobasins and two-ring basins on the Moon. Thus, a spectrum of modification is represented in this collection (Figure 8.9), ranging from the endogenic style of Crisium to the simple, gravity-controlled style of Orientale; the transitional Nectaris, Imbrium, and Serenitatis basins display both styles.

The formation of multi-ring basins is a complex process involving cratering mechanics, volcanism, tectonism, ballistic sedimentation, and endogenic modification. Thus, to speak of basins as "large craters" is a vast oversimplification. The lunar data provide much information on the nature of the basin-forming process and this process possesses subtle complexity, much of it heretofore unappreciated. In the next chapter, I will use the geological model developed here for basins on the Moon to infer the geological processes responsible for the development of multi-ring basins that occur on the other terrestrial planets.

9

Multi-ring basins on the terrestrial planets

The formation of multi-ring basins was an important process in the early histories of Solar System bodies. Thus, study of basins on the other planets potentially can give us insight into the early geological evolution of the planets. Although occurring on all of the terrestrial planets, the most and best preserved basins occur on planets that display remnants of their early crusts, i.e., Mercury, Mars, and the icy satellites of Jupiter and Saturn. In this chapter, I discuss the geology of basins found on the terrestrial planets in relation to the geological model for basin formation and development on the Moon discussed above.

9.1 Earth

Most of the recognized impact structures on the Earth are either simple, bowl-shaped craters or complex craters displaying central peaks (Grieve and Robertson, 1979; Masaitis et al., 1980; Grieve, 1987). However, several of Earth's larger craters have multiple rings; seventeen craters display at least two rings (Table 9.1; Pike, 1985 and references therein). The paucity of terrestrial multi-ring basins doubtless reflects the relatively youthful average surface age of the Earth, as compared with the more primitive terrestrial planets, such as Mercury and Mars.

Impact craters of the Earth show the morphological transitions with increasing size, as do craters on the planets, but changes in form occur at different diameters (Pike, 1985). Complex craters on the Earth range in size from about 4 km to about 25 km in diameter. The smallest crater that has multiple rings, yet possesses a central peak, is Haughton (Robertson and Sweeney, 1983); the largest impact crater on the Earth yet identified is the Vredefort structure of South Africa (Dietz, 1961), or possibly, the Sudbury basin, Canada (Figure 9.1; Grieve et al., 1991). The transitions that mark the onset of protobasin, two-ring, and multi-ring basins on Earth are obscured by the overlap in diameters of these types of features; all three of these morphologies are displayed in the diameter range of 21 km to 150 km (Table 9.1; Pike, 1985). Moreover, target rocks are apparently an important variable in the determination of crater morphology on the Earth (e.g., Pike, 1980c). The six basins

Table 9.1 *Protobasins and multi-ring basins of the Earth*

Crater	Main rim diameter (km)	Other ring diameters (km)
Deep Bay	10	19
Haughton	21	9, 14.5
Clearwater East	22	44
Gosses Bluff	22	11
Mistastin	23	12
Clearwater West	32	16, 64
Charlevoix	55	28
Puchezh-Katunki	80	42
Manicouagan	100	55, 75, 140
Zhamanshin	13	6.5
Wanapitei	17	8.5, 12, 24
Strangways	22.5	10
Ries	25	12, 34, 45
Carswell	37	19
Siljan	52	22
Popigay	95	45, 72, 140
Vredefort	150	75
Sudbury	200 (?)	—
Chicxulub	204	104, 150

Source: After Pike (1985)

in Table 9.1 that exhibit three or more rings apparently follow the same $\sqrt{2}\,D$ spacing rule for adjacent rings that is found on the Moon (Chapter 8) and on Mercury and Mars (Pike, 1985; Pike and Spudis, 1987).

An origin by impact for the Sudbury basin, Canada (Figure 9.1) was first proposed by Dietz (1964), who searched for and discovered shatter cones in the target rocks of the basin. Detailed study of suevitic breccias (the Onaping Formation; French, 1968) and their relation to the large, underlying igneous complex led to the idea that the differentiated igneous rocks of Sudbury basin were the result of impact-triggered magmatic activity (e.g., French, 1970), such as had been often postulated for the Moon. Recently, the geology of the Sudbury basin has been re-examined (Grieve *et al.*, 1991). Detailing the occurrences of dikes of impact breccia, shatter cones, and shock metamorphic features, these authors propose that Sudbury is a much larger impact structure than had been previously believed; the transient crater for Sudbury was on the order of about 100 km, implying an original rim diameter on the order of 200 km (Figure 9.1; Grieve *et al.*, 1991). Thus, Sudbury was well within the morphological range expected for multi-ring impact basins, although it is unclear whether the basin ever displayed multiple rings.

A startling consequence of this new interpretation of the size of the Sudbury basin is the origin of the ore-bearing igneous complex. Because the basin is much larger than supposed, it must have originally possessed a very large volume of impact melt, a much greater volume than can be presently accounted for by masses of melt that occur within the suevitic Onaping Formation (French, 1968). Thus, the igneous complex of Sudbury may be a thick sheet of impact melt that, having contained few entrained clasts, cooled slowly and differentiated in place (Grieve *et al.*, 1991). Such a scenario could have profound implications for our interpretations of the geological record of the planets; if such a process operates in the large melt sheets of basins, many of the lunar samples that we consider to be of endogenous igneous origin (i.e., the so-called pristine rocks; Warren and Wasson, 1977) could be samples of a large, differentiated melt sheet (Grieve *et al.*, 1991). Although I believe that this process is unlikely to be responsible for most of the pristine lunar samples, the concept that large melt sheets might stay molten long enough to differentiate casts doubt on our ideas about the homogeneity of impact melts, especially the melt sheets of basins of increasing size. Differentiated impact melt sheets may be important contributors of "igneous" rock types to early planetary crusts.

The poor preservation of ancient surfaces on the Earth ensures that no large, terrestrial multi-ring basin can furnish a complete model to basins on the planets. However, as noted in Chapter 2, certain aspects of the impact process have been

Figure 9.1 Geological sketch map of the Sudbury basin, Ontario, Canada. The spatial distribution of shock metamorphic features (e.g., shatter cones and mineral deformations) and crater-related geological units (e.g., sub-floor breccia dikes and suevite layers) suggest that this crater may have been as large as 200 km in diameter immediately after formation (its current elliptical shape is caused by regional, Grenville deformation). If this reconstruction is correct, Sudbury would be the largest impact basin yet recognized on Earth. After Grieve *et al.* (1991).

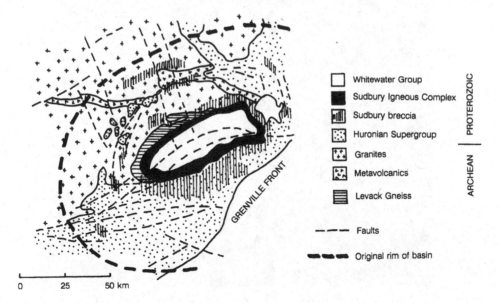

studied profitably by examining those features that are preserved at terrestrial impact craters, such as melt sheets, ejected melt bombs, continuous ejecta, and central peaks and crater structures (see Masaitis *et al.*, 1976). Moreover, the hypothesis of mass extinctions of life caused by large impacts (Alvarez *et al.*, 1980) gives us a reference datum by which to understand the global effects of large-body impact on planetary atmospheres and climate. In this sense, the Earth has provided us with a guide for the geological interpretation of impact craters on the planets.

9.2 Mercury

Mercury is a body very similar to the Moon in surface morphology. Initial examination of photographs of Mercury taken by Mariner 10 revealed many impact features, including large basin-sized structures (Murray *et al.*, 1974). The surface gravity of Mercury is about twice that of the Moon and this difference has some interesting consequences for the crater-forming process; for example, ejecta facies occur closer to the crater rim on Mercury than on the Moon (Gault *et al.*, 1975).

Well-defined protobasins and two-ring basins are abundant on Mercury. Pike

Figure 9.2 The mercurian two-ring basin Bach (69° S, 101° W; 210 km diameter). Inner ring (103 km diameter) is crisp and well-preserved; basin floor is flooded by younger, smooth plains materials. North at top; portion of H-15 photomosaic.

(1988) compiled a detailed inventory of mercurian protobasins and two-ring basins and documented the morphological transitions of impact features on Mercury. A typical two-ring basin is Bach (Figure 9.2); the inner peak ring of Bach is very well developed and remarkably circular. The transition from complex crater to proto-basin occurs in the diameter range between 80 and 200 km (Figure 9.3). Two-ring basins begin to appear around 100 to 200 km in diameter; above about 300 km, basins are multi-ringed, except for those where burial by younger units or degradation have removed traces of additional rings (e.g., Beethoven; Spudis and Guest, 1988). The ring spacing of basins on Mercury is at the same $\sqrt{2}\,D$ interval seen on the Moon (Figure 9.4; Pike and Spudis, 1987). The lithospheric and crustal structure of Mercury probably differed from that of the Moon early in its history, suggesting that these two variables do not dominate the mechanism of basin ring formation, as discussed in Chapter 8.

Figure 9.3 The crater to basin transition on Mercury. Horizontal axis is observed diameter (D_m) of main rim of crater and vertical axis (D_n) are diameters of central peak (shaded), peak plus ring (circle and triangle) for proto-basins, inner ring (dots) for two ring basins, and all mapped rings (crosses) for multi-ring basins. After Pike (1988).

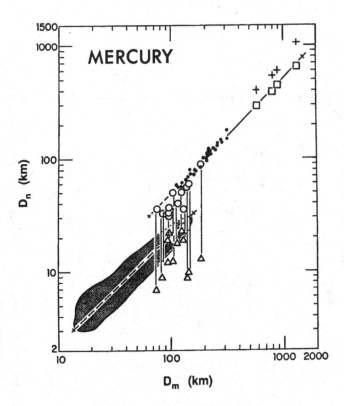

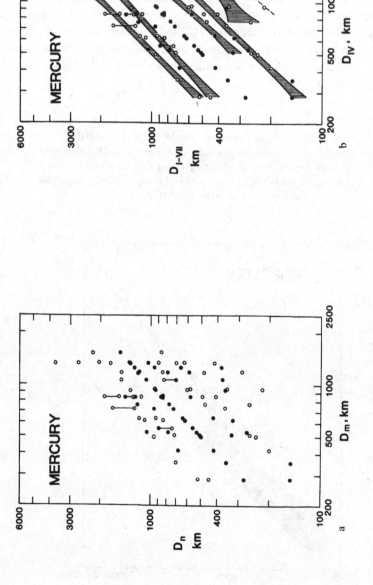

Figure 9.4 Ring spacing for mercurian multi-ring basins (see Figure 8.7). D_m is the observed rim diameter; D_n are all other observed rings. Split rings shown by vertical bars; less reliable rings by open circles. (a) Unranked rings. (b) Ranked rings showing least-squares linear fits and 95% confidence interval for rings other than the main rim (ring IV). Mercurian basin rings are spaced at the $\sqrt{2} D$ interval. From Pike and Spudis (1987).

Table 9.2 *Mercurian multi-ring basins*

No.[a]	Basin[b]	Center	Age[c]	Ring diameters (km)[d]
1	Caloris	30N, 195W	C	630, 900, **1340**, 2050, 2700, 3700
2	Tolstoj	16S, 164W	T	260, 330, **510**, 720
3	Van Eyck	44N, 159W	pT	150, **285**, 450, 520
4	Shakespeare	49N, 151W	pT	200, **420**, 680
5	Sobkou	34N, 132W	pT	490, **850**, 1420
6	Brahms–Zola	59N, 172W	pT	340, **620**, 840, 1080
7	Hiroshige–Mahler	16S, 23W	pT	150, **355**, 700
8	Mena–Theophanes	1S, 129W	pT	260, 475, **770**, 1200
9	Tir	6N, 168W	pT	380, 660, 950, **1250**
10	Andal–Coleridge	43S, 49W	pT	420, 700, 1030, **1300**, 1750
11	Matisse–Repin	24S, 75W	pT	410, **850**, 1250, 1550, 1990
12	Vincente–Yakovlev	52S, 162W	pT	360, **725**, 950, 1250, 1700
13	Eitoku–Milton	23S, 171W	pT	280, 590, 850, **1180**
14	Borealis	73N, 53W	pT	860, **1530**, 2230
15	Derzhavin–Sor Juana	51N, 27W	pT	**560**, 740, 890
16	Budh	17N, 151W	pT	580, **850**, 1140
17	Ibsen–Petrarch	31S, 30W	pT	425, **640**, 930, 1175
18	Hawthorne–Riemenschneider	56S, 105W	pT	270, **500**, 780, 1050
19	Gluck–Holbein	35N, 19W	pT	240, **500**, 950
20	Chong–Gauguin	57N, 106W	pT	220, 350, 580, **940**
21	Donne–Molière	4N, 10W	pT	375, 700, 825, **1060**, 1500
22	Bartók–Ives	33S, 115 W	pT	480, 790, **1175**, 1500
23	Sadi–Scopas	83S, 44W	pT	360, 600, **930**, 1310

[a]Number of basin indicated on map (Figure 9.6)
[b]After Spudis (1983), Pike and Spudis (1987), and Spudis and Guest (1988).
[c]Relative ages: C–Calorian; T–Tolstojan; pT–pre-Tolstojan – see Spudis and Guest (1988).
[d]Diameter in bold face is main topographic rim of basin.

9.2.1 Ancient mercurian basins

In continuing study of Mariner 10 photographs, many workers attempted to identify and inventory the basin population of Mercury (Wood and Head, 1976; Malin, 1976; Schaber *et al.*, 1977; Croft, 1979; Frey and Lowry, 1979). The list of ancient basins recognized in the sparse photographic coverage of Mercury continued to grow; Spudis (1983) compiled a list of all recognized basins on Mercury having multi-ring form (Table 9.2). Many of these basins were discovered in the course of systematic mapping of the global geology of Mercury (Spudis and Guest, 1988).

The ancient multi-ring basins of Mercury were recognized by morphologic criteria similar to those employed in the discovery of degraded basins on the Moon. Such criteria include, (1) arcs of massif chains and isolated massifs that protrude through other units, (2) arcuate lobate ridges (*rupes*) that align with massifs in circular patterns, (3) arcuate highland scarps aligned with the massifs and ridges, and

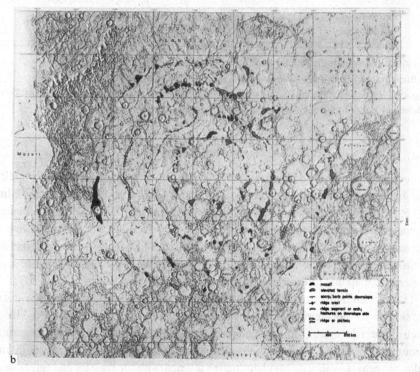

(4) isolated zones of anomalous topographic highs in areas of heavily cratered terrain. Most multi-ring basins on Mercury predate the intercrater plains (Spudis and Guest, 1988), previously held to be the oldest unit on the planet (Trask and Guest, 1975). Because systematic photographic coverage of Mercury is incomplete, the inventory of basins (Table 9.2) is underestimated (Spudis, 1983).

The Tir basin is typical of the ancient basins of Mercury (Figure 9.5). This basin lies within a regional structural trough associated with the Caloris basin (see below) and is filled with smooth plains materials. Abundant massifs (Figure 9.5a,b) align with circular arrangements of mare-type ridges to outline the ring structure of the basin. Mapping topographic and structural elements associated with the basin results in concentric patterns (Figure 9.5b), to which circles are fitted that best represent the ring structure of the basin (Figure 9.5c). Such a procedure was used to compile the basin ring inventory presented in Table 9.2.

Ancient basins are randomly distributed on the hemisphere of Mercury imaged by Mariner 10 (Figure 9.6) and probably represent the remnants of a lunar-like,

Figure 9.5 Series showing the ancient Tir basin, Mercury. (a) Photomosaic of the H-8 quadrangle. (b) Terrain map of basin elements and ridges recognized from mosaic. (c) Best-fit rings to the terrain map shown in (b) See Table 9.2 for locations and ring diameters of all mercurian multi-ring basins. From Pike and Spudis (1987).

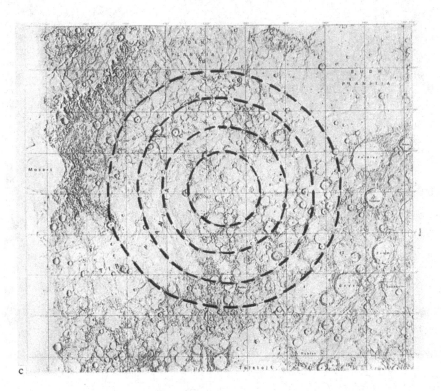

heavily cratered terrain that has been extensively resurfaced by the intercrater plains (Spudis and Guest, 1988). These basins served as depositional sites for the emplacement of the younger smooth plains and the trends of tectonic features, such as ridges, are influenced by the presence of old basin rings. Thus, the basins formed the broad-scale stratigraphic and structural framework for the subsequent geological evolution of Mercury's surface.

9.2.2 The Caloris basin

The most prominent feature identified in the Mariner 10 coverage of Mercury is the Caloris basin (Strom *et al.*, 1975; Trask and Guest, 1975). This basin has been the subject of several detailed geological studies (Strom *et al.*, 1975; McCauley, 1977; McCauley *et al.*, 1981) and will be only briefly described here. Caloris resembles

Figure 9.6 Basins and their ring systems on the hemisphere of Mercury imaged by Mariner 10. Numbers refer to list in Table 9.2. Equal-area projection.

the Imbrium basin of the Moon in many respects, but also has distinct morphology that is the product of the unique mercurian environment.

The Caloris basin is 1340 km in diameter (Figure 9.7) and the basin topographic rim consists of massifs, arranged in a concentric pattern; this ring has been interpreted to be analogous to the Apennine ring of the Imbrium basin (Strom *et al.*, 1975) or to the Outer Rook ring of the Orientale basin (McCauley, 1977). Well

Figure 9.7 The Caloris basin, Mercury. Basin rim (1300 km diameter) is marked by massifs and rough, hummocky basin deposits. Lineated terrain (L) northwest of the rim is ejecta (the Van Eyck Formation, lineated facies). Hummocky and knobby deposits (K) dominate ejecta elsewhere. Basin interior is mostly flooded by smooth plains; concentric texture reflects buried rings. Platform massif at bottom center (arrow). Mariner 10 photomosaic.

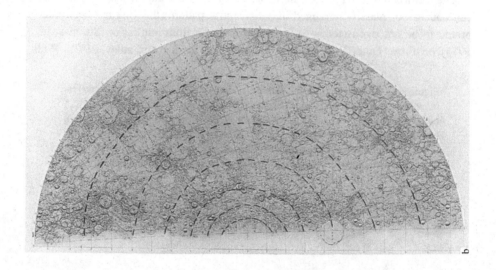

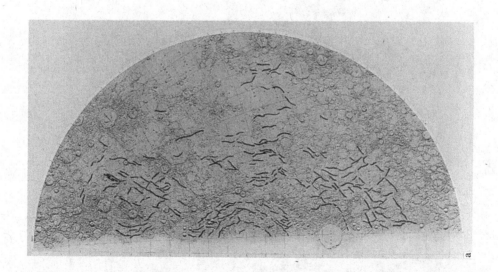

developed textured ejecta are sparse to absent around most of the basin periphery, although lineated materials (the Van Eyck Formation, lineated facies; McCauley *et al.*, 1981) are exposed to the northeast of the basin. An undulating to smooth unit (the Nervo Formation) mantles the intermassif areas near the basin rim and may consist of impact melt, fallback ejecta, or a combination of the two. A knobby unit, similar to the knobby deposits around the Orientale and Imbrium basins (Chapters 3 and 7), is found in the western sectors of the exterior of the Caloris basin (the Odin Formation); this material probably represents some facies of ejecta. A field of secondary craters (Van Eyck Formation; secondary crater facies) was recognized during systematic mapping (McCauley *et al.*, 1981). Distal clusters of these Caloris secondaries are useful in establishing relative ages of basins (e.g., Spudis and Guest, 1988), in a manner similar to the use of secondaries from the Imbrium basin on the Moon (Wilhelms, 1976a, 1987).

The ring system of the Caloris basin is the subject of some controversy. The only obvious ring is the 1340 km main basin rim, defined by the Caloris Montes; a weakly developed outer scarp, equated with the Cordillera scarp of the Orientale basin by McCauley (1977), is visible in the northeast sector of the basin exterior. This scarp is close (about 200 km) to the basin rim and may be an example of a paired or "split" ring fragment (Pike and Spudis, 1987). The basin interior is almost completely flooded by plains materials, probably of volcanic origin as demonstrated by their crater density, which is significantly lower than that for the basin deposits (Spudis and Guest, 1988). The extensive burial of the basin interior of Caloris by plains material has largely removed evidence for inner rings, although the radial and concentric fractures in the central basin are probably related to buried ring structures (Strom *et al.*, 1975; Dzurisin, 1978; Spudis and Guest, 1988). Additionally, a broad, regional pattern of concentric structures is centered around the Caloris basin (Figure 9.8a); this pattern is made up of wrinkle ridges, scarps, and physiographic boundaries both inside and outside of the basin. I interpret the distribution of these structures to mean that the Caloris basin has six rings, ranging in size from 630 km to 3700 km in diameter (Figure 9.8b; Table 9.2).

Despite superficial differences, the Caloris basin bears a striking similarity to the lunar Imbrium basin (Chapter 7). Both have six rings, two inside the rugged massif-lined ring and three exterior rings, defined by ridges and scarps. In addition, both basins exhibit large tracts of smooth, volcanic plains in broad arcs external to the basin rim (Oceanus Procellarum around the Imbrium basin; Tir, Odin, and Susei Planitia around the Caloris basin).

I interpret the rugged, mountainous ring of Caloris as the main topographic rim

Figure 9.8 Basin-centered view of the Caloris basin showing distribution of radial and concentric structures. Stereographic projection by U.S. Geological Survey; north at top. (a) Map of major ridges within Caloris Planitia and smooth plains exterior to basin. (b) Ring interpretation map, based on data shown in (a) See Table 9.2 for ring diameters. From Spudis and Guest (1988). Copyright University of Arizona Press 1988.

of the basin (1340 km diameter) and believe that it is equivalent to the Apennine ring of the Imbrium basin. The poorly expressed scarp evident in the northwest sector of the basin exterior (McCauley, 1977) is a split ring segment (Wilhelms *et al.*, 1977; Pike and Spudis, 1987) and thus, part of the main rim of the basin. By analogy with lunar basins, this interpretation implies that the transient cavity of the Caloris basin was smaller than and lies within the present topographic basin. The absence of rings inside the basin reflects their burial by smooth plains materials of volcanic origin.

Caloris has undergone modification of its morphology by internal processes, including both sub-lithospheric flow and topographic relaxation by viscous flow. Parts of the basin rim are made up of platform massifs (Figure 9.7). Moreover, the basin floor is deformed, showing evidence for first compression (caused by floor subsidence) and then extension (caused by doming of the floor; Dzurisin, 1978). Some of these processes are similar to those that produced the morphology of basins on the Moon, as at Crisium (Chapter 5) and portions of the Imbrium basin rim (Chapter 7). Morphologic differences between lunar basins and the Caloris basin may reflect the higher surface gravity of Mercury, which would concentrate Caloris ejecta closer to the basin rim (Gault *et al.*, 1975). A differing crustal composition and thinner lithosphere at the time of the Caloris impact would also affect the subsequent endogenic modification of basin topography.

9.3 Mars

Mars displays more complex geology than the Moon, has twice the lunar gravitational acceleration, and has a thin atmosphere. The planet has been active volcanically throughout most of its history (Greeley and Spudis, 1981), which implies that it has a relatively thin lithosphere compared to the Moon. Thus, martian basins might be expected to display evidence for internal modification, depending upon both the size of the basin-forming impact and the local conditions of its target site on Mars.

The discovery of basins on Mars closely followed the global mapping of the planet by the Mariner 9 mission in 1971 (Hartmann and Raper, 1974). Several martian basins are morphologically similar to the younger lunar basins. The peak-ring basin Lowell closely resembles the Schrödinger basin on the Moon (Wilhelms, 1973; Wood and Head, 1976). Additionally, protobasins on Mars have been identified (Pike and Spudis, 1987) and the crater-to-basin transitional sequence has been quantified (Figure 9.9).

Few martian basins display pristine morphology because they formed early in the planet's history and the complex and protracted history of Mars has obscured systematic transitions in morphology between basins. The difficulty in determining the transition diameters of martian craters and basins partly reflects the fact that, in general, martian basins tend to be considerably more degraded than those found on the Moon or Mercury (Wilhelms, 1973; Schultz and Glicken, 1978; Schultz *et al.*, 1982). Onset diameters for these morphological types overlap more on Mars than they do on the Moon or Mercury; protobasins appear at around 65 km diameter and persist to diameters approaching 160 km. Martian two-ring basins can be as small as 52 km and as large as 400 km in diameter; this last value is for Schiaparelli, a feature so filled with younger material, that originally it was probably a multi-ring basin (Stam *et al.*, 1984). True multi-ring basins on Mars begin to appear at diameters of about 300 km. The spacing of rings in martian multi-ring basins is shown in Figure 9.10 (Pike and Spudis, 1987). Again, the pervasive $\sqrt{2} D$ spacing rule applies to martian basins as it does for basins on the Moon and Mercury.

Figure 9.9 The crater to basin transition on Mars. Horizontal axis is observed main rim diameter of crater (D_m); vertical axis (D_n) are diameters of central peak complex (shaded), peak plus ring (circle and triangle) for protobasins, inner ring (dots) for two-ring basins, and all rings for multi-ring basins (crosses). After Pike and Spudis (1987).

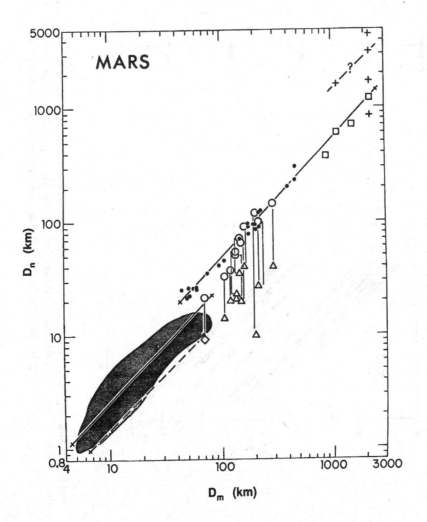

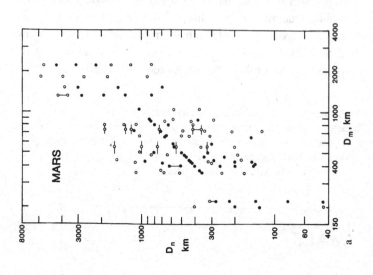

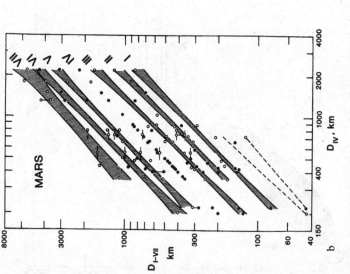

Figure 9.10 Ring spacing for multi-ring basins on Mars. D_m is observed main rim diameter and D_n is diameter of other rings; connected circles are split rings; open circles are less reliable values. (a) Unranked data. (b) Ranked data showing least-squares linear fits with 95% confidence intervals (shaded). As on the Moon and Mercury, martian basins follow the $\sqrt{2}\,D$ spacing rule. From Pike and Spudis (1987).

Table 9.3 *Martian multi-ring basins*

No.[a]	Basin[b]	Center	Ring diameters (km)[c]
1	near Newcombe	22S, 4W	290, 380, 580, **870**
2	Ladon	18S, 29W	325, **550**, 760, 1000, 1600
3	near Holden	25S, 32W	300, **580**
4	Chryse	24N, 45W	840, **1320**, 1920, 2840, 3940
5	Mangala	0 , 147W	500, **660**, 1040
6	Sirenum	44S, 167W	320, **500**, 810, 1040
7	Hephaestus Fosse I	30S, 180W	180, 290, **440**, 1540
8	Al Qahira	20S, 190W	125, 210, 340, 490, **720**, 1140
9	Hephaestus Fosse II	10S, 233W	370, 570, **1040**
10	near Hellas	58S, 275W	200, 360, **550**
11	Nilosyrtis Mensae	32N, 281W	**370**, 720
12	south of Renaudot	38N, 297W	150, 310, 440, **660**
13	south of Lyot	42N, 322W	145, 200, 260, **400**, 565
14	Cassini	24N, 331W	150, 200, 320, **430**, 950
15	Deuteronilus B	43N, 338W	44, 130, **200**, 405
16	Deuteronilus A	44N, 342W	44, 80, 140, **220**, 290
17	near South	73S, 214W	385, **740**, 1205, 1315, 1890
18	near Schiaparelli	5S, 349W	140, 300, **420**, 700, 1120
19	near Le Verrier	37S, 356W	190, **480**, 640, 850
20	Daedalia	26S, 125W	1100, 1500, 2200, **4500**, 6400
21	South Polar	83S, 267W	425, 670, **850**
22	Argyre	50S, 43W	410, 630, **800**, 1100, 1360, 1900
23	Isidis	13N, 272W	700, **1500**, 3000, 3800
24	Hellas	41S, 295W	840, 1200, 1700, **2200**, 3100, 4400, 5500
25	Huygens	14S, 304W	240, 340, **465**
26	near Cassini	14N, 324W	170, 250, **355**, 520, 680, 1100
27	Utopia	48N, 240W	3300, **4715**
28	Elysium	33N, 201W	800, 1540, 2000, **3600**, 4970
29	Hesperia	35S, 248W	1100, **1800**, 2600
—	[North Tharsis]	11N, 97.5W	1455, 2330, 3650, **4500**

[a]Number of basin indicated on map (Figure 9.12).

[b]Basins originally recognized by Schultz and Glicken (1978), Schultz *et al.* (1982), Croft (1979) – South Polar, Stam *et al.* (1984) – near Cassini, Craddock et al. (1990) – Daedalia, Schultz (1984) – Elysium, and R. A. Schultz and Frey (1990) – [North Tharsis]. All ring measurements by Pike and Spudis (1987), except for Daedalia, Utopia, Elysium, and North Tharsis, which are from original references.

[c] Diameter in bold face is main topographic rim of basin.

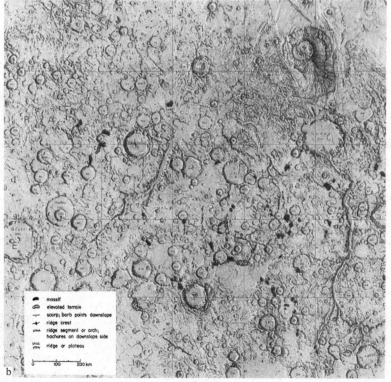

massif

elevated terrain

scarp; barb points downslope

ridge crest

ridge segment or arch;
hachures on downslope side

ridge or plateau

0   100   200 km

9.3.1 Ancient martian basins

Our knowledge of the number and distribution of martian multi-ring basins has been expanded dramatically through the work of Schultz and Glicken (1979), Schultz *et al.* (1982), and R. A. Schultz and Frey (1990). These studies identified over 30 multi-ring basins on Mars, in varying states of preservation. I present my estimate of the current inventory of martian multi-ring basins in Table 9.3; ring diameters are taken from Pike and Spudis (1987) and my own measurements. Although authors may disagree about the precise diameters of basin rings, there is virtual unanimous agreement regarding the *presence* of a given basin on Mars (cf. Table 9.3 with Table 1 of Schultz and Frey, 1990).

An example of an ancient martian basin is Al Qahira (Figure 9.11; Schultz *et al.*, 1982; Pike and Spudis, 1987). This basin, in the southern cratered-terrain hemi-

Figure 9.11 The Al Qahira basin, Mars (Schultz *et al.*, 1982). (a) Photomosaic of part of the MC-23 quadrangle. Large channel at right is Ma'adim Vallis. (b) terrain map of basin features associated with Al Qahira basin; massifs and ridges outline rings. (c) Ring interpretation map, derived from terrain map (b). See Table 9.3 for ring diameters. From Pike and Spudis (1987).

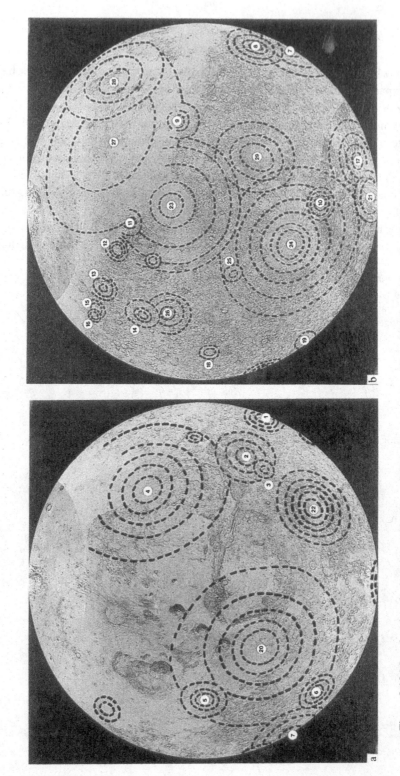

Figure 9.12 Map showing the locations of basins and their ring systems on Mars. Numbers refer to list in Table 9.3. Equal-area projection.

sphere, has been eroded and buried extensively by the superposition of heavily cratered terrain and plains materials; portions of basin deposits also have been incised by channeling. The basin is recognized by abundant massifs protruding through the smooth plains, arcuate scarps and ridges, and the regional trend of a channel, Ma'adim Vallis (Figure 9.11a). In the interpretation of Schultz *et al.* (1982), the deep-seated structures established during the formation of basins control subsequent events, such as the general trend of a channel and its deflection by basin structures (Figure 9.11b). Additionally, topography is particularly pronounced near a massif cluster at 26° S, 185° W that probably relates to another ancient basin centered to the south (Table 9.3; Hephaestus Fosse I; Schultz *et al.*, 1982). This reinforcement and enhancement of topography for intersecting basin rings also occurs on the Moon (e.g., in the Apennines of the Imbrium basin; see Chapter 7) and aids recognition of degraded basins on the planets (Pike and Spudis, 1987).

These ancient multi-ring basins (Figure 9.12) may have had far-reaching effects on the subsequent geological evolution of Mars. Schultz and Glicken (1979) identified several highland massifs protruding through volcanic plains materials south of Noctis Labyrinthus. Detailed mapping of these massifs and the regional tectonic patterns suggested to them that a very large (~1400 km diameter) basin underlies the Noctis Labyrinthus, which is the center of uplift for the entire Tharsis bulge (e.g., Mutch *et al.*, 1976). Schultz and Glicken (1978) proposed that this basin experienced volcano-tectonic modification and served as a locus for the protracted volcanic history of the Tharsis region. Alternatively, Craddock et al. (1990) suggest that highland massifs and localized occurrences of lineated terrain indicate the presence of a large basin centered south of the Tharsis uplift, a basin that they name the Daedalia basin (Figure 9.12, no. 20; Table 9.3). In either case, the presence of a large, multi-ring basin served as the locus for extensive volcanic and tectonic activity. Detailed geological mapping of the Chryse basin likewise delineated the dominant role of the multi-ring basin in the volcanic, fluvial, and tectonic history of that region of Mars (Schultz *et al.*, 1982).

Besides these ancient, nearly obliterated martian basins, several multi-ring structures have been identified that apparently formed at a late stage during the heavy bombardment of Mars. The Isidis basin is extensively flooded by lava (Wilhelms, 1973; Scott and Carr, 1978), but resembles the lunar Serenitatis basin in morphology and preservation state (Chapter 6). Isidis probably formed late in the early bombardment of Mars, when the martian lithosphere was relatively rigid; the morphology of the basin suggests that this impact was not large enough to penetrate the lithosphere. Thus, endogenic modification was of limited extent for these youngest basins on Mars.

Martian basins appear to have undergone considerably more internal and external modification than their lunar counterparts. The paucity of pristine morphologies for martian basins is probably a consequence both of the thin lithosphere in early martian history, which would cause immediate modification to basin morphology, and the continuous volcanic activity over most of geological time, which would bury and modify basin topography. Schultz and Glicken (1978) and Schultz (1984) suggest that basin-forming impacts on Mars initiated extensive volcanism and that long-term, volcano-tectonic modification of basins has been more important on Mars than the Moon. The basin-forming model developed in Chapter 8 supports this concept, and martian basin morphology provides a model and guide for interpreting ancient lunar basins. Thus, some of the older lunar basins (e.g., Tranquillitatis) may have been similar in morphological details to ancient martian basins.

9.3.2 Argyre and Hellas

Argyre is a large, multi-ring basin in the cratered terrain hemisphere of Mars (Wilhelms, 1973). It is similar in appearance to the lunar Crisium basin in that the basin interior and troughs between basin rings have been extensively flooded by volcanic plains materials (Figure 9.13; Wilhelms, 1973; Hodges, 1980); the basin

Figure 9.13 Photomosaic of the Argyre basin, Mars. Clusters of massifs (M) may be martian equivalent of lunar basin platform massifs. Ring troughs (arrows) occur between massif clusters. Basin interior inundated by lava flows and eolian sediments.

has six rings (Table 9.3). Platform topography is not evident; the ring morphology of Argyre appears instead to consist of clusters of rugged massifs (Figure 9.13). These clusters differ from the equant massifs of basin rings on the Moon in that they are not confined narrowly to concentric patterns, such as those of the Outer Rook ring of the Orientale basin (Chapter 3). The massif clusters of the Argyre basin may be the martian equivalent to the platform massifs found around some lunar basins.

The Argyre basin has undergone a history comparable to the lunar Crisium basin. The transient cavity of Argyre may not be represented by any of the currently expressed basin rings; an initially small cavity could have been buried by the thick volcanic fill evident in the interior of the basin. The development of concentric ring troughs (Figure 9.13) suggests that the impact penetrated the martian lithosphere; foundering of crustal blocks and sub-lithospheric flow could have produced both the ring troughs and massif clusters. Flooding of the troughs by lava was immediate and is analogous to the flooding in the interplatform regions of the lunar Crisium basin by ancient mare basalts (sec. 5.2). Ejecta from the Argyre basin cannot be identified because they have been buried by the plateau plains material of probable volcanic origin (Greeley and Spudis, 1981).

Hellas is the largest martian basin that retains continuous rim structure. It has as many as seven rings (Table 9.3), the main rim being defined by the rugged, 2200 km diameter ring of massifs (Pike and Spudis, 1987). The basin is extensively modified and flooded by volcanic plains materials; only a vestige of the basin geology is preserved. In places, the Hellas rings consist of narrow, arcuate massifs that are arranged in concentric patterns (Wilhelms, 1973; Potter, 1976; Peterson, 1977). The intermassif areas are flooded by volcanic materials, some of which are among the oldest plains on Mars (Greeley and Spudis, 1978; 1981). Probably none of the rings of the Hellas basin are related to the transient cavity; this feature, if indeed it was ever expressed, has been totally obliterated by later modification and extensive burial by volcanic units, analogous to the development of the Crisium basin on the Moon.

9.3.3 Polar basins and the hemispheric dichotomy

One of the major discoveries of the exploration of Mars is recognition of a fundamental dichotomy of physiography, *viz.*, the mostly southern hemisphere contains the bulk of the ancient cratered terrain whereas the northern hemisphere is extensively resurfaced by younger units of varied origins (Mutch *et al.*, 1976). Wilhelms and Squyres (1984) suggested that the northern polar region of Mars is underlain by an ancient, gigantic basin ("Borealis basin", 7700 km diameter) that stripped off the cratered terrain in this area, paving the way for continued volcanic resurfacing of this hemisphere throughout martian history. A basin of this size, which subtends almost a third of the martian globe, would be the largest basin in the Solar System. However, Schultz (1984) suggested instead that the Borealis basin actually consists

Table 9.4 *Multi-ring basins on the satellites of Jupiter and Saturn*

Satellite	Basin	Center	Ring diameters (km)
Ganymede	Gilgamesh	59S, 124W	180, 300, 470, **700**, 940, 1200
	near equator	8S, 114W	75, 110, 175, **225**, 305, 440
Callisto	Valhalla	11N, 58W	over 24 rings, up to 1500 km
	Asgard	30N, 140W	over 12 rings, up to 700 km
Tethys	Odysseus	29N, 128W	230, 320, **425**, 610, 780
Rhea	Kun Lun	30N, 307W	460, 650, **900**, 1350
	(unnamed)	35N, 150W	250, **400**, 550
	near Leza	10S, 310W	**350**, 700

Note: Ganymede, Callisto, and Tethys basins discovered by Voyager imaging team (Smith *et al.*, 1979, 1981); Rhea basins discovered in Moore *et al.* (1985) and by the author and R.J. Pike in 1984. All ring diameters my own mapping. Diameter is in boldface topographic rim of basin.

of a series of smaller, coalesced basins in this hemisphere. I believe that at least four basins lie beneath Vastitas Borealis on Mars, including a basin centered near the Elysium volcanic province (Schultz, 1984), a very large basin to the west beneath Utopia Planitia (McGill, 1989), and two or more basins north of Tharsis (Schultz, 1984; R. A. Schultz and Frey, 1990). Moreover, according to Wilhelms and Squyres (1984), part of the rationale for postulating the Borealis basin is analogy with "Procellarum basin" on the Moon; as discussed in Chapter 7, the existence of Procellarum basin on the Moon is suspect and its rings may in fact represent exterior rings of the Imbrium basin.

Regardless of whether the northern hemisphere of Mars is underlain by a single, large basin or by a series of smaller, superposed basins, the evidence for control of the distribution of geological units on planetary surfaces is compelling. Not only are large tracts of plains units preferentially distributed within basin environments of deposition (Schultz and Frey, 1990), but also regional structural patterns, including the development of some of the martian canyons, appear to be related to concentric and radial trends of ancient basins on Mars (Schultz and Glicken, 1978; Schultz, 1984). The hemispheric dichotomy of Mars is the biggest planetary feature yet explained by the far reaching effects of basin-forming impacts.

9.4 Icy satellites of Jupiter and Saturn

9.4.1 Ganymede and Callisto

The two icy Galilean satellites of Jupiter, Ganymede and Callisto, have ancient surfaces which harbor multi-ring basins (Smith *et al.*, 1979). These basins provide important information about the impact process because they formed in a non-silicate target on bodies with low gravity. Thus, some of the morphological differences evident in these basins may help toward understanding some of the peculiarities of basin morphology on silicate planets.

Basins on the surface of Ganymede are similar to those on the Moon (Shoemaker *et al.*, 1982). Both peak-ring and multi-ring basins are present; the best example of a multi-ring basin is Gilgamesh (Figure 9.14). Gilgamesh displays at least four, and possibly as many as six, rings (Table 9.4); it also has a textured ejecta blanket, secondary craters up to 1000 km from the basin rim, and an interior filled by smooth plains materials (Shoemaker *et al.*, 1982). Its rim (700 km diameter) is composed of rectilinear segments similar to *en echelon* normal faults, broadly arranged into a circular pattern (Figure 9.14). One other basin on Ganymede (Table 9.4) is similar to the smaller multi-ring basins on Mercury.

In addition to these "normal" impact basins, the large, dark area Galileo Regio

Figure 9.14 The Gilgamesh basin, Ganymede (see Table 9.4). Basin with textured ejecta (E) appears similar to those on rocky planets; concentric scarps (arrows) show *en echelon* offsets, suggesting structural origin. Interior flooded by smooth plains deposits, probably caused by water ice volcanism (Shoemaker *et al.*, 1982). Voyager 2 image 0527J2-001.

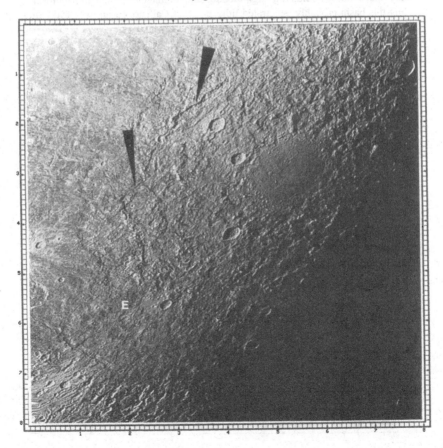

appears to be the site of a highly degraded, ancient basin (Shoemaker *et al.*, 1982). This dark region of relatively smooth terrain is surrounded by an arcuate band of grooved materials that probably make up the rim materials of a highly modified, ancient basin. The smaller basins of Ganymede formed in a thick, rigid lithosphere; subsequent internal modification of their topography has been minimal. In contrast, the Galileo Regio basin formed very early in the planet's history (McKinnon and Melosh, 1980) and has been modified by global tectonism and ice volcanism (Shoemaker *et al.*, 1982).

On Callisto, the spectacular Valhalla structure demonstrates that multi-ring basins may assume many diverse morphologies (Figure 9.15). This feature contains more than two dozen concentric ring arcs; no single arc makes up a complete ring and no single arc is topographically dominant. The zone of ring arcs extends out to over 2000 km from the basin center (McKinnon and Melosh, 1980) and a smooth, high-albedo region (about 600 km in diameter) occupies the basin center. In detail, the Valhalla basin rings resemble simple scarps and in many places are polygonal

Figure 9.15 The Valhalla basin, Callisto. Basin displays over 24 concentric ring arcs surrounding a relatively high-albedo, smooth inner zone. The impact forming this basin probably penetrated lithosphere of Callisto. Basin over 1500 km in diameter. Voyager 1 image 1761J1+000.

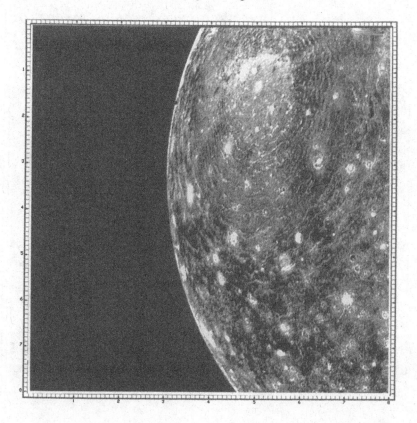

in outline (Figure 9.15). Some of the inter-ring regions appear to be "flooded" by smooth plains material (McKinnon and Melosh, 1980), similar to relations seen in some highly modified lunar and martian basins (secs. 5.2, 9.3). The smooth interior area of the basin may be caused by "volcanic" flooding of melted ices and clathrates (Smith *et al.*, 1979).

The Valhalla and Asgard basins of Callisto do *not* conform to the ring spacing relation ($\sqrt{2} D$) observed on every other body of the Solar System. Because they are unique in this, and other, aspects, it has even been suggested that they are not of impact origin, but were formed instead by internal processes (e.g., Wood, 1981). Evidence from the basins of Callisto suggests that the process of lithospheric penetration by impact can be of great importance in the evolution of basin morphology. The unusual morphology of the Valhalla basin suggests that penetration of a very thin lithosphere may produce many rings, probably through the process of gravity-controlled slumping of the crust around a small cavity (Melosh and McKinnon, 1978; McKinnon and Melosh, 1980).

9.4.2 Saturnian satellites

The small, icy satellites of Saturn were observed during the Voyager encounters to be covered with impact craters (Smith *et al.*, 1981), but the recognition of multiring basins on Rhea and Tethys took detailed photogeological mapping and study (Moore and Horner, 1984; Moore *et al.*, 1984; Moore *et al.*, 1985). To date, four multi-ring structures have been recognized on the Saturnian satellites (Table 9.4).

Tethys is a small (525 km radius) icy body that has a prominent impact feature, Odysseus. Independent mapping of basin rings by R.J. Pike and the author in 1984 demonstrated that Odysseus consists of at least three, and possibly as many as five, rings (Table 9.4). The central peak complex of Odysseus indicates that it is not a conventional "multi-ring" basin, but may be a protobasin. The multiple rings mapped around Odysseus may reflect fracturing and relaxation caused by the penetration of the Tethys lithosphere when this feature formed early in the moon's history.

The Saturnian satellite Rhea (765 km radius) has three basins (Moore *et al.*, 1985) of a much more "terrestrial" planet character (Table 9.4; Figure 9.16). These basins are very ancient and nearly obliterated by superposed craters. Basin rings are evident by arcuate scarps and lineaments concentric to a central point. A poorly photographed basin on the anti-Saturn hemisphere of Rhea appears much more regular in form; it is probably the youngest of the three features. Moore *et al.* (1985) suggest that the basins on Rhea were important sites for the control of planetary resurfacing or mantling; the least heavily cratered terrain on Rhea occurs within these basins. Thus, the pattern of basin control of geological provinces seen on the inner planets appears to hold for this satellite, possessor of the most distant basins from the Sun yet recognized.

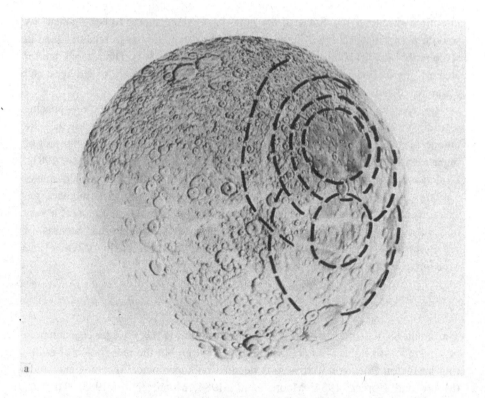

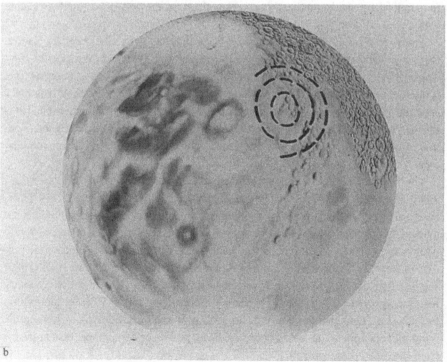

9.5 Venus

The detailed morphology of the surface of Venus has become known through the results of the successful Magellan mission (Saunders *et al.*, 1991). Additionally, the Soviet Venera 15 and 16 spacecraft mapped about 25 percent of Venus with side-looking radar in 1983 and 1984 (Barsukov *et al.*, 1986). On the basis of these data and some selected high-resolution images obtained from Earth-based radar (Campbell *et al.*, 1984), it appears that Venus totally lacks the ancient, heavily cratered terrain that occurs on Mercury, Mars, and the Moon. Venus is similar to the Earth in that active volcanism and tectonism have acted in concert to produce a relatively young surface; the average surface age of Venus is currently estimated to be between 250 Ma and 450 Ma old, the younger age being more likely (Schaber *et al.*, 1987). Thus, impact craters in general are relatively rare on Venus and the larger ones, which include basins, are exceedingly uncommon. However, several impact basins have been identified within the Magellan coverage (Phillips *et al.*, 1991; Schaber *et al.*, 1992) and other features of ambiguous morphology ultimately may prove to be of impact origin.

The crater Klenova (Figure 9.17), near Tethus Regio, appears to be a true multi-ring basin (Barsukov *et al.*, 1986; Ivanov *et al.*, 1986; Schaber *et al.*, 1992). It has two discontinuous but well-defined rings of about 150 and 105 km diameter and a smaller, more ill-defined ring about 70 km in diameter. Portions of the main, outer ring split into small segments of arcuate, subparallel scarps, similar to the split rings of some other planetary basins, such as Caloris (sec. 9.2). Klenova has a relatively bright radar halo, 40 to 60 km wide, indicating a rough surface, probably the ejecta blanket of the crater. The floor appears to be relatively smooth, with some zones of knobby terrain close to the inner ring (Ivanov *et al.*, 1986).

Typical planetary two-ring basins show ring spacing of 2:1 for the outer/inner rings (Pike and Spudis, 1987). The ratio of the ring diameters for Klenova is about 1.4:1; on the other planets, this spacing describes *multi-ring* basins. Pike and Spudis (1987) argue that the predominant spacings for two-ring and multi-ring basins are actually part of the same general spacing rule for all basin rings, i.e., ring *locations* are spaced at integer powers $(\sqrt{2})^n D$, where D is the diameter of the ring lying inside of an observed ring and n is an integer. Thus, Klenova is a true multi-ring basin, not a two-ring basin. Additional rings may be buried or weakly developed; multiple rings of impact structures on the Earth also can be very subtle in topographic expression (e.g., Manicouagan; Floran and Dence, 1976). If Klenova is a multi-ring basin, it falls into the diameter range in which multiple rings are devel-

Figure 9.16 Shaded relief map of Saturnian satellite Rhea, showing location of three multi-ring basins described by Moore *et al.* (1985). Relatively smooth terrain is confined to these basin interiors, suggesting early resurfacing of Rhea was basin-controlled. Map by U.S. Geological Survey.

oped on Earth; it is comparable in size to the terrestrial Vredefort structure. Schaber *et al.* (1992) found that the $\sqrt{2}$ spacing rule held for multi-ring basins on Venus.

The Magellan coverage has revealed many additional impact basins (Figure 9.18), most of which display at least two (sometimes three) rings and display near pristine morphology (Schaber *et al.*, 1992). Mona Lisa and Stanton (Figure 9.18) are multi-ring and two-ring basins filled with dark and light plains materials, respectively. Both basins have blocky rim ejecta, distributed in lobes or ray-like patterns. The slightly elliptical Isabella (Figure 9.18c) is a multi-ring basin with dark fill material and an extremely long outflow of presumed impact melt (Schaber

Figure 9.17 The crater Klenova (78° N, 104° E; main rim - 150 km diameter) on Venus, imaged by the *Magellan* spacecraft. Inner rings (70 and 105 km in diameter) consist of blocky massif elements and ridged elements, some of which are offset from center. Smoothed dark area within basin probably post-impact lava flooding. Blocky ejecta evident outside basin rim, some of which are concentrated in "ray" deposits. Spacing of rings in Klenova ($\sqrt{2}\,D$) suggests that it is a true "multi-ring" basin, similar to those observed on the other terrestrial planets.

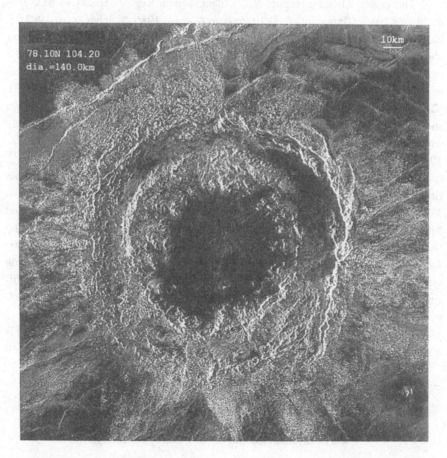

78.10N 104.20
dia.=140.0km

10km

For legend see page 222.

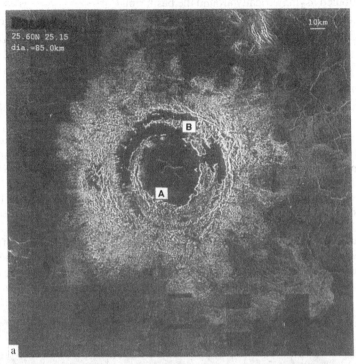

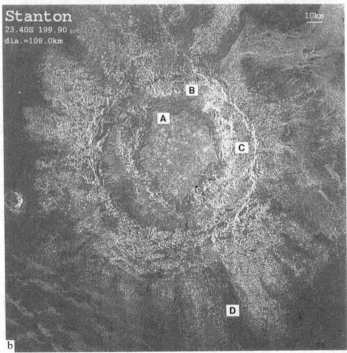

et al., 1992). This basin may be the result of oblique impact, as its exterior ejecta are distributed asymmetrically around the basin rim.

The total number of impact craters on Venus revealed by the Magellan coverage is meager, but those that are found fall into morphological classes and size ranges similar to those on the Earth. These observations, and the relative crater density of Venus, suggest that the cratering records of Earth and Venus are very similar. Because the most ancient terrain of Venus does not date from the era of heavy bombardment (Schaber *et al.*, 1987; Phillips *et al.*, 1992), the basin population of

Figure 9.18 Typical multi-ring basins on Venus, as seen from *Magellan*; north at top in each. (a) Mona Lisa (85 km dia.) has radar-dark floor materials. Inner ring (A) is more complete than the intermediate ring (B). Ejecta deposits extend outward about a crater radius from the basin rim. (b) Stanton (108 km dia.) has radar-bright floor with dark "apron" deposits (A) along the inner base of the peak ring. Traces of the intermediate ring are evident in the north (B) and east (C). "Rays" of basin ejecta appear in several sectors exterior to the rim (D). (c) Isabella (165 km dia.) is one of the largest basins yet identified on Venus. The scarp-like eastern rim (A) contrasts with the hummocky western and southern rim material (B). Central region is deeply flooded and these deposits display many scarps and ridges, some of which may reflect subsurface ring structure.

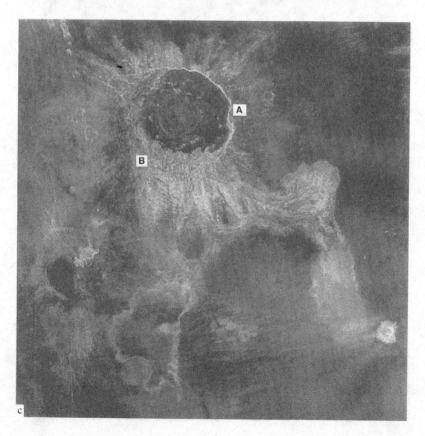

Venus is dominated by the smallest members of that category. Moreover, Venus basins may provide a unique reference for the interpretation of small basins on the Earth, as both planets have nearly the same gravity and, to a lesser degree, comparable targets (e.g., basaltic rocks with high thermal gradients) and environments (e.g., presence of an atmosphere). Venus basins may permit us to reconstruct the original morphology and topography of highly degraded terrestrial impact basins, such as Sudbury and Vredefort. Conversely, the geology of these terrestrial basins tell us about the likely structural and petrological effects associated with basin formation on Venus. Thus, the basins of Venus and Earth complement each other by providing interpretive guides for the geological elements that are missing on each planet.

10

Multi-ring basins and planetary evolution

The formation of multi-ring basins is one of the most important geological processes in the early history of the Solar System. These impacts can greatly affect the morphology and observed surface composition of planetary crusts. The formation of basins may influence lithospheric development and growth and thus alter the thermal history of the planet. Basin-forming impacts can catalyze volcanic eruptions and initiate major modifications of crustal structure subsequent to the development of their multi-ring topography.

In this chapter, I will conclude my examination of the geology of multi-ring basins by speculating on the role of basins in the early geological evolution of the planets. Many of the ideas offered in this chapter are subjects of ongoing research and answers to some of the questions raised by such speculation may be forthcoming with additional work.

10.1 The building blocks of planetary surfaces

The recognition of regional patterns of landforms on the Moon led to the discovery of multi-ring basins; this pattern recognition of "the big picture" out of the chaos of detail displayed by the lunar surface is well described by Hartmann (1981). The use of such perception techniques in planetary photogeology has shown us that basins are also present on Mercury, Mars, and the icy satellites of the jovian planets. Moreover, such discovery is not yet complete; ongoing analysis of planetary images adds every year to the basin inventory of the Solar System.

From the evidence described and analyzed in this book, I believe that basins are the fundamental building blocks of early, planetary crusts. Initial impressions of chaos displayed in heavily cratered terrains on the terrestrial planets have gradually given way to perceptions of order. In his history of the U.S. Geological Survey's program to map the Moon geologically, Wilhelms (1990, 1993) describes the initial collective bewilderment as geologists attempted to extend mapping from the relatively simple maria into the complex highlands. Early *ad hoc* interpretations of volcanic origins for the units of the highlands gradually gave way to recognition that virtually all highland landforms and geological units are of impact origin. This shift in paradigm was largely a result of study of the spectacular Orientale basin on the Moon, where landforms and surface textures are relatively fresh. Study of the geology of Orientale basin permitted the extension of impact interpretations to other highland areas, where erosional degradation made geological relations more obscure.

Even with the lunar experience behind us, ten years passed before it became apparent that basins, especially old basins, were no less important on Mercury and Mars than they are on the Moon. For example, isolated and aligned massifs in the cratered terrain of Mars were recognized soon after the Viking Orbiter images were received; many of these features were interpreted to be volcanic constructs (e.g., Greeley and Spudis, 1978; Scott and Tanaka, 1981). It took painstaking mapping of regional patterns to show that, in fact, most of these features are parts of ancient, degraded multi-ring basins (Schultz and Glicken, 1978; Schultz *et al.*, 1982). We now recognize that basins, originally thought to be deficient on Mars and Mercury relative to the Moon (Wood and Head, 1976), make up the structural and geological framework of these planets.

10.2 Effects of basins on planetary evolution

Basins have profoundly affected the subsequent geological evolution of all the planets and satellites studied to date. Although such effects take slightly different form on different planetary bodies, the most important geological processes attributable to basins are volcanism and tectonism.

10.2.1 The Moon

It is no coincidence that Baldwin (1949) originally applied the term "circular maria" to the multi-ring basins on the Moon. Of the 16 percent of the lunar surface covered by the visible maria (Head, 1976b), virtually all mare deposits are basin-contained or basin-related. This correlation is not surprising; mare basalts form by the partial melting of the lunar mantle and merely require structural paths through the crust to reach the surface. Large impact basins not only excavate millions of cubic kilometers of crust, but also induce major fractures in the lithosphere. Basins provide conduits for the delivery of magma to the surface, topographic lows for the accumulation of lava flows, and topographic depressions to contain them. It is likely that some basins actually induce volcanism, as has been postulated for the origin of the KREEP basalts (the Apennine Bench Formation) of the Imbrium basin (Spudis, 1978; Ryder 1988).

In addition to localizing volcanic activity, basins have dominated lunar structural patterns. The regional trends of compressive wrinkle ridges in the maria delineate the ring systems of many mare-filled basins. The loading of the lunar lithosphere by dense stacks of mare basalt flows deforms the crust, creating compressive, interior ridges and peripheral, extensional graben (Solomon and Head, 1980). Radial and concentric fractures produced during basin-forming impacts leave a structural imprint over vast regions of the Moon (Mason *et al.*, 1976). In short, the Moon would be a very different place geologically without its population of large, multi-ring basins.

Basins on the Moon serve as loci for a variety of geological activity, including crust-forming magmatism. Multi-color data from the flyby of the *Galileo* spacecraft the Moon (Belton *et al.*, 1992) show an albedo and color anomaly associated with the floor of the giant South Pole–Aitken basin (Figure 10.1; Wilhelms, 1987). This zone of low albedo has spectral properties that suggest it is more mafic than the typical surrounding highlands (Belton *et al.*, 1992); its boundaries also coincide with a zone classified as Mg-suite on the petrologic province map of Davis and Spudis (1987). Together, these observations indicate that a large Mg-suite pluton occurs in this region of the Moon; the pluton could either have been exposed by the

Figure 10.1 Western limb and far side of the Moon as seen by the *Galileo* spacecraft in December, 1990. The dark region on the left limb of the Moon is the zone of mafic deposits that are confined within the South Pole–Aitken basin. This localization of composition is strongly suggestive of basin control of the distribution of petrological units in the highlands crust.

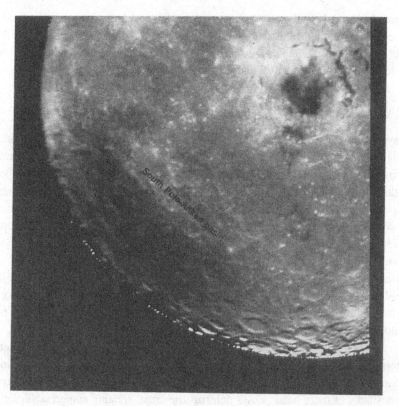

impact that formed South Pole–Aitken or that impact could have triggered the igneous activity. In either case, a large basin has significantly affected the observed composition of a region of the Moon.

10.2.2 Mercury

The striking Caloris basin and its deposits have dominated discussion of the effects of basin-forming impacts on Mercury (e.g., Strom, 1984). However, the large, ancient basins beneath the inter-crater plains (Spudis, 1983) have significantly influenced the observed distribution of geological units and tectonic features on Mercury and these obscure structures are arguably much more important to the configuration of the planet than Caloris.

Figure 10.2 Basins control the distribution of volcanic units on Mercury, as they do on the Moon. This map shows the distribution of smooth plains (shaded) and the locations of ancient, multi-ring basins on Mercury. Note that the locations of the deposits of smooth plains are strongly correlated with the presence of a basin. Lambert equal-area projection, after Spudis and Guest (1988). Copyright University of Arizona Press 1988.

Smooth Plains
(Calorian-Tolstojan)

It was initially thought that the mercurian smooth plains were largely concentrated within and near the Caloris basin (Trask and Guest, 1975); indeed, this supposed distribution led some to question the volcanic origin of the smooth plains of Mercury (Wilhelms, 1976b). However, subsequent geological mapping revealed two important points: (1) the smooth plains in fact have a widespread distribution over the entire hemisphere of Mercury imaged by Mariner 10; and (2) these smooth plains tend to be concentrated within and near the sites of ancient multi-ring basins, which cannot be responsible for them because the plains are of Calorian age (Figure 10.2; Spudis and Guest, 1988). These observations, and the younger ages of the plains relative to the deposits of Caloris basin (Spudis and Guest, 1988), indicate that the plains are volcanic lavas, analogous to the lunar maria. Moreover, subsequent tectonism, expressed partly by compressive thrust faults that are widely distributed over the surface of Mercury (Melosh and McKinnon, 1988), show circular patterns coincident with ancient basin rings; indeed, this relation was one used by Spudis (1983) to discover some of these features. Thus, the observed pattern of geological and tectonic units on Mercury strongly reflects the presence of multi-ring basins.

10.2.3 Mars

The formation of basins on Mars may have initiated the development of the Tharsis volcanic complex and the hemispheric dichotomy (e.g., Schultz and Glicken, 1978; Wilhelms and Squyres, 1984). Abundant lava plains on the planet occupy depressions created by basins. Major martian volcanoes may owe their locations to the intersections of basin radial and concentric structures (e.g., Peterson, 1978; Schultz, 1984). The youngest four major basins on Mars (Argyre, Hellas, Chryse, and Isidis) have had important influences on the distribution of geological units (Schultz, 1984); the development of chaotic terrain, and even the canyon of Vallis Marineris, follow pre-existing structural trends that may be related to the Chryse basin. Basins in the Tharsis region (Figure 9.11) established a structural pattern for the massive flooding of the martian surface by lavas, and could be responsible for the development of a thermal plume in the mantle that ultimately led to the Tharsis construct. Flooding of the martian surface by lava plains has obscured many of these geological relations, but global mapping of wrinkle ridges in the plains of Mars outline, patterns radial and concentric to mapped, ancient basins (Chicarro and Schultz, 1982).

The origin of the hemispheric dichotomy on Mars is an even more profound consequence of basin impact. I consider this controversial idea to be well founded, regardless of whether the dichotomy results from several superposed basins (Schultz, 1984; McGill, 1989; Schultz and Frey, 1990) or from the giant Borealis basin of Wilhelms and Squyres (1984). Abundant massifs are found along the margins of the terrain boundary between cratered uplands and smooth lowlands on Mars. Endogenic interpretations of the hemispheric dichotomy (e.g., Mutch *et al.*,

1976) are still entertained; while I do not question the importance of such processes in the development of the current configuration of the planet, the evidence for multi-ring basins as the ultimate cause of the dichotomy seems very compelling.

10.2.4 Icy satellites of Jupiter and Saturn

On Ganymede, the circular arrangement of grooved terrain around the dark area of Galileo Regio led to the hypothesis that this region was the site of an ancient impact basin (Shoemaker *et al.*, 1982). I accept this interpretation, and further suggest that additional ancient basins, currently unrecognizable in the poor photographic coverage of Ganymede provided by the Voyager missions, formed the nuclei around which all of the grooved terrain formed. This speculation can be tested when global, high-resolution photographs of Ganymede are returned from the Galileo mission. The youngest basins on Ganymede (e.g., Gilgamesh) formed when the lithosphere had grown too thick to be penetrated by an impact. These features have served to localize extrusions of water-ice volcanism, as evident from the deposits of smooth plains contained in their interiors.

Callisto, Tethys, and Rhea are all primitive bodies, yet evidence for resurfacing events in their history shows that while such activity is meager, it occurs within the confines of multi-ring basins. The interiors of Valhalla and Asgard contain the least heavily cratered terrain on Callisto; the basin ring systems here outline the "active" zones of an otherwise primitive, cratered surface (Figure 9.15). On Rhea, similar positioning of relatively smooth geological units suggests that the three basins on that body (Figure 9.16) likewise provide catalysts for internal processes. Future missions to these and other satellites of the outer planets may yet reveal additional effects of ancient basins on the primitive, icy bodies of the Solar System.

10.2.5 Earth

Because the other terrestrial planets experienced basin-forming impacts during the heavy bombardment, there is no reason to suppose that Earth escaped the effects of these catastrophic events. Two separate models propose that basins on the early Earth had major geological consequences. Frey (1980) suggested that large basins forming on the Earth 4 Ga ago eventually led to the creation of Earth's ocean basins; in contrast, Grieve (1980b) suggested that such impact sites would become centers of intensive igneous activity, ultimately leading to the production of terrestrial proto-continents (Figure 10.3). I believe that Grieve's scenario is the more likely of the two. We have seen ample evidence on the other planets for intensive igneous activity associated with the sites of multi-ring basins; on the early Earth, such activity would have occurred in the presence of abundant free water, which aids in the production of intermediate and granitic magmas (Campbell and Taylor, 1983). Grieve (1980b) suggests that basin-forming impacts on the Earth led to the formation of proto-continental nuclei in the early Archean; over the subsequent 3.8

billion years of Earth history, material accreted onto these nuclei to produce the continents in their present configuration.

Such a consideration provides food for thought for the future exploration of Mars; because this planet apparently was water-rich early in its history (Carr, 1986) and evidence for igneous activity related to basins is abundant (Schultz and Glicken, 1978; Schultz, 1984), igneous complexes having petrologic similarities with terrestrial proto-continents may exist beneath the floors of some ancient martian basins. This speculation can be tested during future exploration of Mars by examining the ejecta of later craters forming on these basin floors, that would bring such material to the surface.

Basin impacts on Earth later in the Precambrian could have had different consequences. These impacts may have initiated early continental rifting and ultimately could have been responsible for triggering plate tectonics. Ironically, such initiation would be responsible for later removing the evidence for their former existence.

Figure 10.3 Possible effects of large basins on the early (Archean) Earth. Schematic cross sections show initial disturbance (top) causes mantle upwelling. This change, a consequence of excavation of crustal materials, results in pressure-release melting and copious volcanism. Basin subsides (middle), collecting sediments, but is still very active volcanically. Reprocessing of crater fill (bottom) results in generation of tonalitic magmas, which rise diapirically. Continued igneous activity at depth generates granitic magmas, resulting in a complex igneous body, possibly a protocontinental nucleus. After Grieve (1980b).

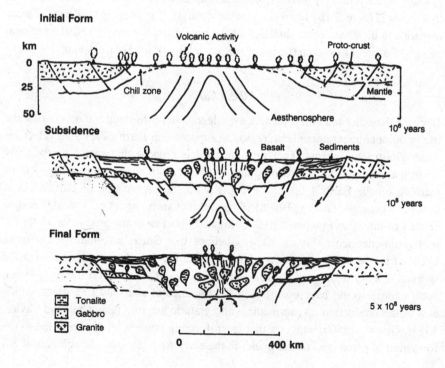

The formation of multi-ring basins on Earth may have far-reaching ramifications that are not immediately apparent, except by the study of a number of seemingly unrelated phenomena. An example of increasing importance is the problem of the mass extinction of species at the end of the Cretaceous Period (e.g., Silver and Schultz, 1982; Alvarez, 1987). The discovery of the worldwide iridium anomaly at the K–T boundary has dramatically focused attention on the effects of large impacts; Alvarez *et al.* (1980) calculated that the impact at the end of the Cretaceous was large enough to form a crater over 100 km in diameter. Thus, the basin problem once again becomes important, this time, for the history and evolution of life itself, as originally suggested by Baldwin (1949).

Recently, a significant amount of evidence suggests that the Chicxulub structure of northern Yucatan, Mexico, may be the long-sought crater that initiated the Cretaceous–Tertiary extinction event (Hildebrand *et al.*, 1991). Chicxulub crater is about 200 km in diameter and is buried by a thick sequence of Tertiary carbonates; as defined by gravity anomalies, it has at least two rings of 100 and 150 km diameter inside the main rim (Hildebrand *et al.*, 1991). Radiometric dating of impact melt rocks contained within suevitic breccias indicates that the crater formed at the end of the Cretaceous Period, about 65 Ma ago (Swisher *et al.*, 1992; Sharpton *et al.*, 1992). The worldwide effects of the Chicxulub impact have been extensively documented (e.g., Sharpton and Ward, 1990) and included the extinction of over 90 percent of all living species. Study of the formation Chicxulub crater and its consequences provides us with a guide for the interpretation of the global atmospheric and climatic effects of basin formation, a topic of great significance in the formation and geological effects of basins on Mars and Venus.

10.3 Conclusion

Planets are diverse and complex objects. The interaction of these bodies with a physical process as enigmatic as a basin-forming impact presents planetary geologists with an intriguing problem. I have attempted to unravel some of the complexity surrounding this problem by showing how basins on the Moon, our nearest planetary neighbor, can elucidate the processes of excavation and modification that make up basin formation. The solutions that I offer here are but precursors to those that address all of the problems associated with multi-ring basins. Complete understanding of multi-ring basins requires many more years of study.

References

Adler I. and Trombka J.I. (1977) Orbital chemistry–lunar surface analysis from the X-ray and gamma-ray remote sensing experiments. *Phys. Chem. Earth* **10**, 17–43.

Apollo Field Geology Investigation Team (AFGIT) (1973) Geologic exploration of Taurus–Littrow: Apollo 17 landing site. *Science* **182**, 672–680.

Albee A.L., Gancarz A.J., and Chodes A.A. (1973) Metamorphism of Apollo 16 and 17 and Luna 20 metaclastic rocks at about 3.95 AE: Samples 61156, 64423, 14–2, 65015, 67483, 15–2, 76055, 22006, and 22007. *Proc. Lunar Sci. Conf.* **4**, 569–595.

Alvarez L.W. (1987) Mass extinctions caused by large bolide impacts. *Physics Today* **40**, 24–33.

Alvarez L.W., Alvarez W., Asaro F., and Michel H.V. (1980) Extraterrestrial cause for the Cretaceous-Tertiary extinction. *Science* **208**, 1095–1107.

Andre C.G. and El-Baz F. (1981) Regional chemical setting at the Apollo 16 landing site and the importance of the Kant Plateau. *Proc. Lunar Planet. Sci.* **12B**, 767–779.

Alexander E.C. and Kahl S.B. (1974) ^{40}Ar–^{39}Ar studies of lunar breccias. *Proc. Lunar Sci. Cont.* **5**, 1353–1373.

Austin M.G. and Hawke B.R. (1981) Tentative speculations on lunar sample transport. In *Lunar and Planetary Science* **XII**, pp. 34–36. Lunar and Planetary Institute, Houston.

Baldwin R.B. (1949) *The Face of the Moon.* Univ. Chicago Press, Chicago, 239 pp.

Baldwin R.B. (1963) *The Measure of the Moon.* Univ. of Chicago Press, Chicago, 488 pp.

Baldwin R.B. (1974) On the origin of the mare basins. *Proc. Lunar Sci. Conf.* **5**, 1–10.

Baldwin R.B. (1978) An overview of impact cratering. *Meteoritics* **13**, 364–379.

Baldwin R.B. (1981) On the tsunami theory of the origin of multi-ring basins. In *Multi-ring Basins* (P.H. Schultz and R.B. Merrill, editors), *Proc. Lunar Planet. Sci.* **12A**, 275–288.

Baldwin R.B. (1987a) On the relative and absolute ages of seven lunar front face basins. I. From viscosity arguments. *Icarus* **71**, 1–18.

Baldwin R.B. (1987b) On the relative and absolute ages of seven lunar front face basins. II. From crater counts. *Icarus* **71**, 19–29.

Barsukov V.L., and 29 others (1986) The geology and geomorphology of Venus surface as revealed by the radar images obtained by Veneras 15 and 16. *Proc. Lunar Planet. Sci. Conf.* **16**, *J. Geophys. Res.* **91**, D378–D398.

Basilevsky A.T., Ivanov B.A., Florensky K.P., Yakovlev O.I., Feldman V.I., Gronovsky L.V., and Sandovsky M.A. (1983) *Udarnyie krateri na lunye i planetakh.* (Impact Craters on the Moon and Planets), Nauka Press, Moscow, 200 pp. (in Russian)

Belton M.J.S. and 18 others (1992) Lunar impact basins and crustal heterogeneity: New western limb and far side data from Galileo. *Science* **255**, 570–576.

Bielefeld M.J. (1977) Lunar surface chemistry of regions common to the orbital X-ray and gamma-ray experiments. *Proc. Lunar Sci. Conf.* **8**, 1131–1147.

Bielefeld M.J., Reedy R.C., Metzger A.E., Trombka J.I. and Arnold J.R. (1976) Surface chemistry of selected lunar regions. *Proc. Lunar Sci. Conf.* **7**, 2661–2676.

Bielefeld M.J., Wildey R.L., and Trombka J.I. (1978) Correlation of chemistry with normal albedo in the Crisium region. In *Mare Crisium: The View from Luna 24* (R.B. Merrill and J.J. Papike, editors), pp. 33–42, Pergamon Press, N.Y.

Bills B.G. and Ferrari A.J. (1976) Lunar crustal thickness. *Proc. Lunar Sci. Conf.* **7**, frontispiece.

Blanchard D.P., Haskin L.A., Jacobs J.W., Brannon J.C., and Korotev R.L. (1975) Major and trace element chemistry of boulder 1 at station 2, Apollo 17. *The Moon* **14**, 359–371.

Boon J.D. and Albritton C.C. (1936) Meteorite craters and their possible relationship to "cryptovolcanic structures". *Field and Laboratory* **5**, 1–9.

Boyce J.M. (1976) Ages of flow units in the lunar near side maria based on Lunar Orbiter IV photographs. *Proc. Lunar Sci. Conf.* **7**, 2717–2728.

Bratt S.R., Solomon S.C., Head J.W., and Thurber C.H. (1985) The deep structure of lunar basins: Implications for basin formation and modification. *J. Geophys. Res.* **90**, 3049–3064.

Bryan W.B., Finger L.W., and Chayes F. (1969) Estimating proportions in petrographic mixture equations by least-squares approximation. *Science* **163**, 926–927.

Cadogan P.H. (1974) Oldest and largest lunar basin? *Nature* **250**, 315–316.

Cameron K.L., Papike J.J., Bence A.E., and Sueno S. (1973) Petrology of fine-grained rock fragments and petrologic implications of single crystals from the Luna 20 soil. *Geochim. Cosmochim. Acta* **37**, 755–793.

Campbell D.B., Head J.W., Harmon J.K., and Hine A.A. (1984) Venus volcanism and rift formation in Beta Regio. *Science* **226**, 167–170.

Campbell I.E. and Taylor S.R. (1983) No water, no granites- no oceans, no continents. *Geophys. Res. Letters* **10**, 1061–1064.

Carlson R.W. and Lugmair G.W. (1979) Sm-Nd constraints on early lunar differentiation and the evolution of KREEP. *Earth Planet. Science Letters* **45**, 123–132.

Carr M.H. (1986) Mars: A water rich planet? *Icarus* **68**, 187–216.

Carr M.H. and El-Baz F. (1971) Geologic map of the Apennine-Hadley region of the Moon (Apollo 15 pre-mission map). *U.S. Geol. Survey Map* **I–723** (sheet 1).

Carr M.H., Crumpler L.S., Cutts J.A., Greeley R., Guest J.E., and Masursky H. (1977) Martian impact craters and emplacement of ejecta by surface flow. *J. Geophys. Res.* **82**, 4055–4065.

Casella C.J. and Binder A.B. (1972) Geologic map of the Cleomedes quadrangle of the Moon. *U.S. Geol. Survey Map* **I–707**.

Chadderton L.T., Krajenbrink F.G., Katz R., and Poveda A. (1969) Standing waves on the Moon. *Nature* **223**, 259–263.

Chao E.C.T. (1974) Impact cratering models and their application to lunar studies - A geologist's view. *Proc. Lunar Sci. Conf.* **5**, 35–52.

Chao E.C.T., Hodges C.A., Boyce J.M., and Soderblom L.A. (1975) Origin of lunar light plains. *J. Res. U.S. Geol. Survey* **3**, 379–392.

Charette M.P., Taylor S.R., Adams J.B., and McCord T.B. (1977) The detection of soils of Fra Mauro basalt and anorthositic gabbro composition in the lunar highlands by remote spectral reflectance techniques. *Proc. Lunar Sci. Conf.* **8**, 1049–1061.

Chicarro A. and Schultz P.H. (1982) Ridges in the old terrains of Mars. In *Lunar and Planetary Science* **XIII**, pp. 88–89, The Lunar and Planetary Institute, Houston.

Cintala M.J., Wood C.A., and Head J.W. (1977) The effects of target characteristics on fresh crater morphology: Preliminary results for the Moon and Mercury. *Proc. Lunar Sci. Conf.* **8**, 3409–3426.

Clark P.E. and Hawke B.R. (1981) Compositional variation in the Hadley Apennine region. *Proc. Lunar Planet. Sci.* **12B**, 727–749.

Craddock R.A., Greeley R., and Christensen P.R. (1990) Evidence for an ancient impact basin in Daedalia Planum, Mars. *J. Geophys. Res.* **95**, 10729–10741.

Croft S.K. (1979) *Impact craters from centimeters to megameters.* Ph.D. dissertation, Univ. California Los Angeles, 264 pp.

Croft S.K. (1980) Cratering flow fields: Implications for the excavation and transient expansion stages of crater formation. *Proc. Lunar Planet. Sci. Conf.* **11**, 2347–2378.

Croft S.K. (1981a) The excavation stage of basin formation: A qualitative model. In *Multi-ring Basins* (P.H. Schultz and R.B. Merrill, editors), *Proc. Lunar Planet. Sci.* **12A**, 207–225.

Croft S.K. (1981b) The modification stage of basin formation: Conditions of ring formation. In *Multi-ring Basins* (P.H. Schultz and R.B. Merrill, editors), *Proc. Lunar Planet. Sci.* **12A**, 227–257.

Croft S.K. (1985) The scaling of complex craters. *Proc. Lunar Planet. Sci. Conf.* **15**, *J. Geophys. Res.* **90**, C828–C842.

Dalrymple G.B. and Ryder G. (1991) ^{40}Ar/^{39}Ar ages of six Apollo 15 impact melt rocks by laser step heating. *Geophys. Res. Letters* **18**, 1163–1166.

Davis P.A. (1980) Iron and titanium distribution on the moon from orbital gamma-ray spectrometry with implications for crustal evolutionary models. *J. Geophys. Res.* **85**, 3209–3224.

Davis P.A. and Spudis P.D. (1985) Petrologic province maps of the lunar highlands derived from orbital geochemical data. *Proc. Lunar Planet. Sci. Conf.* **16**, *J. Geophys. Res.* **90**, D61–D74.

Davis P.A. and Spudis P.D. (1987) Global petrologic variations on the Moon: A ternary-diagram approach. *Proc. Lunar Planet. Sci. Conf.* **17**, *J. Geophys. Res.* **92**, E387–E395.

De Hon R.A. (1979) Thickness of the western mare basalts. *Proc. Lunar Planet. Sci. Conf.* **10**, 2935–2955.

Delano J.W. (1979) Apollo 15 green glass: Chemistry and possible origin. *Proc. Lunar Planet. Sci. Conf.* **10**, 275–300.

Delano J.W., Bence A.E., Papike J.J., and Cameron K.L. (1973) Petrology of the 2–4 mm fraction from the Descartes region of the Moon and stratigraphic implications. *Proc. Lunar Sci. Conf.* **4**, 537–551.

Dence M.R. (1976) Notes toward an impact model for the Imbrium basin. *Interdisciplinary Studies by the Imbrium Consortium* **1**, Lunar Science Institute Contr. **267D**, 147–155.

Dence M.R. (1977a) The contribution of major impact processes to lunar crustal evolution. *Phil. Trans. Royal Soc. London* **A285**, 259–265.

Dence M.R. (1977b) The Manicouagan impact structure observed from Skylab. In *Skylab Explores the Earth*, NASA **SP-380**, pp. 179–189.

Dence M.R., Grieve R.A.F., and Plant A.G. (1974) The Imbrium basin and its ejecta. In *Lunar Science V*, pp. 165–167. Lunar Science Institute, Houston.

Dence M.R., Grieve R.A.F., and Plant A.G. (1976) Apollo 17 grey breccias and crustal composition in the Serenitatis basin region. *Proc. Lunar Sci. Conf.* **7**, 1821–1832.

Deutsch A. and Stöffler D. (1987) Rb–Sr analyses of Apollo 16 melt rocks and a new age estimate for the Imbrium basin: Lunar basin chronology and the early heavy bombardment of the Moon. *Geochim. Cosmochim. Acta* **51**, 1951–1964.

Dietz R.S. (1946) The meteoritic origin of the Moon's surface features. *J. Geology* **54**, 359–375.

Dietz R.S. (1961) Vredefort ring structure: Meteorite impact scar? *J. Geology* **69**, 499–516.

Dietz R.S. (1964) Sudbury structure as an astrobleme. *J. Geology* **72**, 412–434.

Dowty E., Keil K., Prinz M., Gros J., and Takahashi H. (1976) Meteorite-free Apollo 15 crystalline KREEP. *Proc. Lunar Sci. Conf.* **7**, 1833–1844.

Duncan A.R., Grieve R.A.F., and Weill D.F. (1975) The life and times of Big Bertha: lunar breccia 14321. *Geochim. Cosmochim. Acta* **39**, 265–273.

Dymek R.F., Albee A.L., and Chodos A.A. (1974) Glass-coated soil breccia 15205: Selenologic history and petrologic constraints on the nature of its source region. *Proc. Lunar Sci. Conf.* **5**, 235–260.

Dymek R.F., Albee A.L., and Chodos A.A. (1975) Comparative petrology of lunar cumulate rocks of possible primary origin: Dunite 72415, troctolite 76535, norite 78235, and anorthosite 62237. *Proc. Lunar Sci. Conf.* **6**, 301–341.

Dymek R.F., Albee A.L., and Chodos A.A. (1976) Petrology and origin of Boulders #2 and #3, Apollo 17, Station 2. *Proc. Lunar Sci. Conf.* **7**, 2335–2378.

Dzurisin D. (1978) The tectonic and volcanic history of Mercury as inferred from studies of scarps, ridges, troughs and other lineaments. *J. Geophys. Res.* **83**, 4883–4906.

Eggleton R.E. (1964) Preliminary geology of the Riphaeus quadrangle of the Moon and definition of the Fra Mauro Formation. In *Astrogeol. Studies Ann. Prog. Report*, part A, U.S. Geol. Survey open-file report, p. 46–63.

Eggleton R.E. (1970) Geologic map of the Fra Mauro region of the Moon. *U.S. Geol. Survey Map* **I–708** (sheet 1).

Eggleton R.E. (1981) Map of the impact geology of the Imbrium basin of the Moon. *In* Geology of the Apollo 16 Area-Central lunar highlands. *U.S. Geol. Survey Prof. Paper* **1048**, plate 12.

Eggleton R.E. and Schaber G.G. (1972) Cayley Formation interpreted as basin ejecta. In *Apollo 16 Prelim. Science Rpt.*, NASA **SP–315**, p. 29–7 to 29–16.

Eichhorn G., McGee J.J., James O.B., and Schaeffer O.A. (1979) Consortium breccia 73255: Laser ^{39}Ar-^{40}Ar dating of aphanite samples. *Proc. Lunar Planet. Sci. Conf.* **10**, 763–788.

Englehardt W.v. (1967) Chemical composition of Ries glass bombs. *Geochim. Cosmochim. Acta* **31**, 1677–1689.

Ferrari A.J., Nelson D.L., Sjogren W.L., and Phillips R.J. (1978) The isostatic state of the lunar Apennines and regional surroundings. *J. Geophys. Res.* **83**, 2863–2871.

Fielder G. (1963) Nature of lunar maria. *Nature* **198**, 1256–1260.

Floran R.J. and Dence M.R. (1976) Morphology of the Manicouagan ring-structure, Quebec, and some comparisons with lunar basins and craters. *Proc. Lunar Sci. Conf.* **7**, 2845–2865.

Floran R.J., Grieve R.A.F., Phinney W.C., Warner J.L., Simonds C.H., Blanchard D.P., and Dence M.R. (1978) Manicouagan impact melt, Quebec, 1, Stratigraphy, petrology and chemistry. *J. Geophys. Res.* **83**, 2737–2759.

French B.M. (1968) Sudbury structure, Ontario: Some petrographic evidence for an origin by meteorite impact. In *Shock Metamorphism of Natural Materials* (B.M. French and N.M. Short, editors), pp. 383–412, Mono Book Corp., Baltimore.

French B.M. (1970) Possible relations between meteorite impact and igneous petrogenesis, as indicated by the Sudbury structure, Ontario, Canada. *Bull. Volcanology* **34**(2), 466–517.

Frey H. (1980) Crustal evolution of the early Earth: The role of major impacts. *Precambrian Research* **10**, 195–216.

Frey H. and Lowry B.L. (1979) Large impact basins on Mercury and relative crater production rates. *Proc. Lunar Planet. Sci. Conf.* **10**, 2669–2687.

Gall H.v., Huttner R., and Muller D. (1977) 4. Stratigraphie. Bavarian Geologisches Landesamt, Erläuterungen zur geologische Karte von Bayern, 1:50,000. *Geologica Bavaria* **76**.

Gault D.E. (1974) Impact cratering. In *A Primer In Lunar Geology* (R.Greeley and P. Schultz, editors), pp. 137–175, NASA TM X–**62359**.

Gault D.E., Shoemaker E.M., and Moore H.J. (1963) Spray ejected from the lunar surface by meteoroid impact. *NASA Tech. Note* D–**1767**, 39 pp.

Gault D.E., Quaide W.L., and Oberbeck V.R. (1968) Impact cratering mechanics and structures. In *Shock Metamorphism of Natural Materials* (B.M. French and N.M. Short, editors), pp. 87–99, Mono Book Corp., Baltimore.

Gault D.E., Guest J.E., Murray J.B., Dzurisin D., and Malin M.C. (1975) Some comparisons of impact craters on Mercury and the Moon. *J. Geophys. Res.* **80**, 2444–2460.

Gilbert G.K. (1893) The Moon's face. *Phil. Soc. Wash.* **12**, 241–292.

Gilbert G.K. (1896) The origin of hypotheses, illustrated by the discussion of a topographic problem. *Science, new series,* **3**, 1–13.

Goins N.R., Toksöz M.N., and Dainty A.M. (1979) The lunar interior: A summary report. *Proc. Lunar Planet. Sci. Conf.* **10**, 2421–2439.

Greeley R. (1976) Modes of emplacement of basalt terrains and an analysis of mare volcanism in the Orientale basin. *Proc. Lunar Sci. Conf.* **7**, 2747–2759.

Greeley R. and Spudis P.D. (1978) Volcanism in the cratered terrain hemisphere of Mars. *Geophys. Res. Letters* **5**, 453–455.

Greeley R. and Spudis P.D. (1981) Volcanism on Mars. *Rev. Geophys. Space Phys.* **19**, 13–41.

Greeley R., Fink J., and Gault D.E. (1980) Impact basins: Implications for formation from experiments. *Conf. Multi-ring basins: Their Formation and Evolution,* LPI Contr. 414, pp. 18–20, The Lunar and Planetary Institute, Houston.

Grieve R.A.F. (1975) Petrology and chemistry of the impact melt at Mistastin Lake crater, Labrador. *Bull. Geol. Soc. America* **86**, 1617–1629.

Grieve R.A.F. (1980a) Cratering in the lunar highlands: Some problems with the process, record, and effects. In *Proc. Conf. Lunar Highlands Crust* (J.J. Papike and R.B. Merrill, editors), pp. 173–196, Pergamon Press, N.Y.

Grieve R.A.F. (1980b) Impact bombardment and its role in proto-continental growth on the early Earth. *Precambrian Research* **10**, 217–247.

Grieve R.A.F. (1987) Terrestrial impact structures. *Ann. Rev. Earth Planet. Sci.* **15**, 245–270.

Grieve R.A.F. and Floran R.J. (1978) Manicouagan impact melt, Quebec 2. Chemical interactions with basement and formational processes. *J. Geophys. Res.* **83**, 2761–2771.

Grieve R.A.F. and Garvin J.B. (1984) A geometric model for excavation and modification at terrestrial simple craters. *J. Geophys. Res.* **89**, 11561–11572.

Grieve R.A.F. and Head J.W. (1983) The Manicouagan impact structure: An analysis of its original dimensions and form. *Proc. Lunar Planet. Sci. Conf.* **13**, *J. Geophys. Res.* **88**, A807–A818.

Grieve R.A.F. and Robertson P.B. (1979) The terrestrial cratering record: I. Current status of observations. *Icarus* **38**, 211–229.

Grieve R.A.F., McKay G.A., Smith H.D., and Weill D.F. (1975) Lunar polymict breccia 14321: A petrographic study. *Geochim. Cosmochim. Acta* **39**, 229–245.

Grieve R.A.F., Dence M.R., and Robertson P.B. (1977) Cratering processes: As interpreted from the occurrence of impact melts. In *Impact and Explosion Cratering* (D.J. Roddy, R.O. Pepin, and R.B. Merrill, editors), pp. 791–814, Pergamon Press, N.Y.

Grieve R.A.F., Robertson P.B., and Dence M.R. (1981) Constraints on the formation of ring impact structures, based on terrestrial data. In *Multi-ring Basins* (P.H. Schultz and R.B. Merrill, editors), *Proc. Lunar Planet. Sci.* **12A**, 37–57.

Grieve R.A.F., Stöffler D., and Deutsch A. (1991) The Sudbury structure: Controversial or misunderstood? *J. Geophys. Res.* **96**, 22753–22764.

Gruithuisen F. v. P. (1829) *Analeckten Erd-und-Himmels Kunde.* Munich.

Guest J.E. and Greeley R. (1977) *Geology on the Moon.* Wykeham Publications, London, 235 pp.

Hackman R.J. (1966) Geologic map of the Montes Apenninus region of the Moon. *U.S. Geol. Survey Map* **I–463**.

Haines E.L. and Metzger A.E. (1980) Lunar highland crustal models based on iron concentrations: Isostasy and center-of-mass displacement. *Proc. Lunar Planet. Sci. Conf.* **11**, 689–718.

Hale W.S. and Grieve R.A.F. (1982) Volumetric analysis of complex lunar craters: Implications for basin ring formation. *Proc. Lunar Planet. Sci. Conf.* **13**, *J. Geophys. Res.* **87**, A65–A76.

Hartmann W.K. (1981) Discovery of multi-ring basins: Gestalt perception in planetary science. In *Multi-ring Basins* (P.H. Schultz and R.B. Merrill, editors), *Proc. Lunar Planet. Sci.* **12A**, 79–90.

Hartmann W.K. and Kuiper G.P. (1962) Concentric structures surrounding lunar basins. *Commun. Lunar and Planetary Lab* **1**, Univ. Arizona, Tucson, 55–66.

Hartmann W.K. and Raper O. (1974) *The New Mars: The Discoveries of Mariner 9.* NASA **SP-337**, 179 pp.

Hartmann W.K. and Wood C.A. (1971) Moon: Origin and evolution of multi-ring basins. *The Moon* **3**, 3–78.

Hawke B.R. (1978) Chemical mixing model studies of selected lunar regions. In *Lunar and Planetary Science* **IX**, p. 474–476. Lunar and Planetary Institute, Houston.

Hawke B.R. and Bell J.F. (1981) Remote sensing studies of lunar dark halo craters: Preliminary results and implications for early volcanism. *Proc. Lunar Planet. Sci.* **12B**, 665–678.

Hawke B.R. and Head J.W. (1977a) Impact melt on lunar crater rims. In *Impact and*

Explosion Cratering (D.J. Roddy, R.O. Pepin, and R.B. Merrill, editors), pp.815–841, Pergamon Press, N.Y.

Hawke B.R. and Head J.W. (1977b) Pre-Imbrian history of the Fra Mauro region and Apollo 14 sample provenance. *Proc. Lunar Sci. Conf.* **8**, 2741–2761.

Hawke B.R. and Head J.W. (1978) Lunar KREEP volcanism: geologic evidence for history and mode of emplacement. *Proc. Lunar Planet. Sci. Conf.* **9**, 3285–3309.

Hawke B.R. and Spudis P.D. (1979) Chemical mixing model studies of lunar basin ejecta deposits. In *Papers Presented to the Conference on the Lunar Highlands Crust*, pp. 53–55. Lunar and Planetary Institute, Houston.

Hawke B.R. and Spudis P.D. (1980) Geochemical anomalies on the eastern limb and farside of the Moon. *Proc. Conf. Lunar Highlands Crust* (J.J. Papike and R.B. Merrill, editors), pp. 467–481, Pergamon Press, N.Y.

Hawke B.R., Spudis P.D., and Metzger A.E. (1980) Lunar basin ejecta deposit compositions: A summary of chemical mixing model studies. In *Papers Presented to the Conference on Multi-ring Basins: Formation and Evolution*, pp. 42–44, Lunar and Planetary Institute, Houston.

Hawke B.R., Lucey P.G., Taylor G.J., Bell J.F., Peterson C.A., Blewitt D.T., Horton K., and Spudis P.D. (1991) Remote sensing studies of the Orientale region of the Moon: A pre-Galileo view. *Geophys. Res. Letters* **18**, 2141–2144.

Head J.W. (1974a) Orientale multi-ringed basin interior and implications for the petrogenesis of lunar highland samples. *The Moon* **11**, 327–356.

Head J.W. (1974b) Stratigraphy of the Descartes region (Apollo 16): Implications for the origins of samples. *The Moon* **11**, 77–99.

Head J.W. (1974c) Morphology and structure of the Taurus-Littrow Highlands (Apollo 17): Evidence for their origin and evolution. *The Moon* **9**, 355–395.

Head J.W. (1976a) Evidence for the sedimentary origin of Imbrium sculpture and lunar basin radial texture. *The Moon* **15**, 455–462.

Head J.W. (1976b) Lunar volcanism in space and time. *Rev. Geophys. Space Phys.* **14**, 265–300.

Head J.W. (1977a) Origin of outer rings in lunar multi-ringed basins: Evidence from morphology and ring spacing. In *Impact and Explosion Cratering* (D.J. Roddy, R.O. Pepin and R.B. Merrill, editors), pp. 563–573, Pergamon Press, N.Y.

Head J.W. (1977b) Regional distribution of Imbrium basin deposits: Relationship to pre-Imbrian topography and mode of emplacement. *Interdisciplinary Studies by the Imbrium Consortium* **2**, Lunar Science Institute Contr. **268D**, 120–125.

Head J.W. (1979) Serenitatis multi-ringed basin: Regional geology and basin ring interpretation. *The Moon and Planets* **21**, 439–462.

Head J.W. and Hawke B.R. (1975) Geology of the Apollo 14 region (Fra Mauro): Stratigraphic history and sample provenance. *Proc. Lunar Sci. Conf.* **6**, 2483–2501.

Head J.W. and Hawke B.R. (1981) Geology of the Apollo 16–Descartes region: Stratigraphic history and sample provenance. In *Workshop on Apollo 16* (O.B. James and F. Hörz, editors), pp. 47–49, LPI Tech. Rpt. **81–01**, Lunar and Planetary Institute, Houston.

Head J.W., Adams J.B., McCord T.B., Pieters C., and Zisk S. (1978) Regional stratigraphy and geologic history of Mare Crisium. In *Mare Crisium: The View from Luna 24* (R.B. Merrill and J.J. Papike, editors), pp. 43–74, Pergamon Press, N.Y.

Head J.W., Settle M., and Stein R.S. (1975) Volume of material ejected from major basins and implications for the depth of excavation of lunar samples. *Proc. Lunar Sci. Conf.* **6**, 2805–2829.

Heiken G. and McEwen M.C. (1972) The geologic setting of the Luna 20 site. *Earth Planet. Sci. Letters* **17**, 3–6.

Heiken G.H., McKay D.S., and Brown R.W. (1974) Lunar deposits of possible pyroclastic origin. *Geochim. Cosmochim. Acta* **38**, 1703–1718.

Heiken G., Vaniman D., and French B., editors (1991) *The Lunar Sourcebook*. Lunar and Planetary Institute and Cambridge Univ. Press, N.Y., 736 pp.

Herzberg C.T. (1978) The bearing of spinel cataclasites on the crust–mantle structure of the Moon. *Proc. Lunar Planet. Sci. Conf.* **9**, 319–336.

Hildebrand A.R., Penfield G.T., Kring D.A., Pilkington M., Camargo A., Jacobsen S.B., and Boynton W.V. (1991) Chicxulub crater: A possible Cretaceous/Tertiary boundary impact crater on the Yucatan Peninsula, Mexico. *Geology* **19**, 867–871.

Hodges C.A. (1972) Geologic maps of the Descartes region of the Moon (Apollo 16 pre-mission maps). *U.S. Geol. Survey Map* **I–748** (sheet 2).

Hodges C.A. (1980) Geologic map of the Argyre quadrangle of Mars. *U.S. Geol. Survey Map* **I–1181**.

Hodges C.A. and Muehlberger W.R. (1981) Summary and critique of geologic hypotheses. In Geology of the Apollo 16 Area, Central lunar highlands. *U.S. Geol. Survey Prof. Paper* **1048**, pp. 215–230.

Hodges C.A. and Wilhelms D.E. (1978) Formation of lunar basin rings. *Icarus* **34**, 294–323.

Hood L.L. (1980) Bulk magnetization properties of the Fra Mauro and Reiner Gamma Formations. *Proc. Lunar Planet. Sci. Conf.* **11**, 1879–1896.

Hörz F. and Banholzer G.S. (1980) Deep seated target materials in the continuous deposits of the Ries crater, Germany. *Proc. Conf. Lunar Highlands Crust* (J.J. Papike and R.B. Merrill, editors), pp. 211–231, Pergamon Press, N.Y.

Hörz F., Ostertag R., and Rainey D.A. (1983) Bunte breccia of the Ries: Continuous deposits of large impact craters. *Rev. Geophys. Space Phys.* **21**, 1667–1725.

Howard K.A. (1971) Geologic map of the Apollo 15 landing site (Apollo 15 pre-mission map). *U.S. Geol. Survey Map* **I–723** (sheet 2).

Howard K.A. (1975) Geologic map of the crater Copernicus. *U.S. Geol. Survey Map* **I–840**.

Howard K.A. and Masursky H. (1968) Geologic map of the Ptolemaeus quadrangle of the Moon. *U.S. Geol. Survey Map* **I–566**.

Howard K.A. and Wilshire H.G. (1975) Flows of impact melt at lunar craters. *J. Res. U.S. Geol. Survey* **3**, 237–257.

Howard K.A., Wilhelms D.E., and Scott D.H. (1974) Lunar basin formation and highland stratigraphy. *Rev. Geophys. Space Phys.* **12**, 309–327.

Hoyt W.G. (1987) *Coon Mountain Controversies*. Univ. of Arizona Press, Tucson, 442 pp.

Hubbard N.J. and Minear J.W. (1975) A physical and chemical model of early lunar history. *Proc. Lunar Sci. Conf.* **6**, 1057–1085.

Hubbard N.J., Gast P.W., Rhodes J.M., Bansal B.M., Wiesmann H., and Church S. E. (1972) Nonmare basalts: Part II. *Proc. Lunar Sci. Conf.* **3**, 1161–1179.

Husain L., Schaeffer O.A., and Sutter J.F. (1972) Age of a lunar anorthosite. *Science* **175**, 428–430.

References

Irving A.J. (1977) Chemical variation and fractionation of KREEP basalt magmas. *Proc. Lunar Sci. Conf.* **8**, 2433–2448.

Ivanov B.A., Basilevsky A.T., Kryuchkov V.P., and Chernaya I.M. (1986) Impact craters of Venus: Analysis of Venera 15 and 16 data. *Proc. Lunar Planet. Sci. Conf.* **16**, *J. Geophys. Res.* **91**, D413–D430.

Jackson E.D., Sutton R.L., and Wilshire H.G. (1975) Structure and petrology of a cumulus norite boulder sampled by Apollo 17 in Taurus-Littrow valley, the Moon. *Bull. Geol. Soc. America* **86**, 433–442.

James O.B. (1972) Lunar anorthosite 15415: Texture, mineralogy and metamorphic history. *Science* **175**, 432–436.

James O.B. (1973) Crystallization history of lunar feldspathic basalt 14310. *U.S. Geol. Survey Prof. Paper* **841**, 29 pp.

James O.B. (1980) Rocks of the early lunar crust. *Proc. Lunar Planet. Sci. Conf.* **11**, 365–393.

James O.B. (1981) Petrologic and age relations of the Apollo 16 rocks: Implications for subsurface geology and the age of the Nectaris basin. *Proc. Lunar Planet. Sci.* **12B**, 209–233.

James O.B. and Flohr M.K. (1983) Subdivision of the Mg-suite noritic rocks into Mg-gabbronorites and Mg-norites. *Proc. Lunar Planet. Sci. Conf.* **13**, *J. Geophys. Res.* **88**, A603–A614.

James O.B. and Hörz F., editors (1981) *Workshop on Apollo 16*. Lunar and Planetary Institute Tech. Report **81-01**, 157 pp.

James O.B., Hedenquist J.W., Blanchard D.P., Budahn J.R., and Compston W. (1978) Consortium breccia 73255: Petrology, major- and trace-element chemistry, and Rb–Sr systematics of aphanitic lithologies. *Proc. Lunar Planet. Sci. Conf.* **9**, 789–819.

James O.B., Flohr M.K., and Lindstrom M.M. (1984) Petrology and geochemistry of lunar dimict breccia 61015. *Proc. Lunar Planet. Sci. Conf.* **15**, *J. Geophys. Res.* **89**, C63–C86.

Keil K., Warner R.D., Dowty E., and Prinz M. (1975) Rocks 60618 and 65785: Evidence for admixture of KREEP in lunar impact melts. *Geophys. Res. Letters* **2**, 369–372.

Kieffer S.W. and Simonds C.H. (1980) The role of volatiles and lithology in the impact cratering process. *Rev. Geophys. Space Phys.* **18**, 143–181.

King J.S. and Scott D.H. (1978) The significance of buried craters associated with basins on the Moon and Mars. In *Reports of Planetary Geology Program 1977–78*, p. 153–156, NASA **TM-79729**.

Korotev R.L. and Haskin L.A. (1988) Compositional survey of particles from the Luna 20 regolith. *Lunar Planetary Science* **XIX**, 635–636.

La Jolla Consortium (1977) Global maps of lunar geochemical, geophysical and geologic variables. *Proc. Lunar Sci. Conf.* **8**, frontispiece.

Lance R.H. and Onat E.T. (1962) A comparison of experiments and theory in the plastic bending of circular plates. *J. Mech. Phys. Solids* **10**, 301–311.

Langseth M.G., Keihm S.J., and Peters, K. (1976) Revised lunar heat-flow values. *Proc. Lunar. Sci. Conf.* **7**, 3143–3171.

Laul J.C. and Schmitt R.A. (1973) Chemical composition of Luna 20 rocks and soil and Apollo 16 soil. *Geochim. Cosmochim. Acta* **37**, 927–942.

Leich D.A., Kahl S.B., Kirschbaum A.R., Niemeyer S., and Phinney D. (1975) Rare gas constraints on the history of Boulder 1, Station 2, Apollo 17. *The Moon* **14**, 407–444.

Longhi J. (1978) Pyroxene stability and the composition of the lunar magma ocean. *Proc. Lunar Planet. Sci. Conf.* **9**, 285–306.

Lucchitta B.K. (1972) Geologic sketch map of the candidate Proclus Apollo landing site. In *Apollo 15 Prelim. Sci. Rpt.*, NASA **SP–289**, pp. 25–76 to 25–80.

Lucchitta B.K. (1978) Geologic map of the north side of the Moon. *U.S. Geol. Survey Map* **I–1062**.

Lunar Sample Preliminary Examination Team (LSPET) (1973) Apollo 17 lunar samples: Chemical and petrographic description. *Science* **182**, 659–672.

Mackin J.H. (1969) Origin of lunar maria. *Bull. Geol. Soc. America* **80**, 735–748.

Malin M.C. (1976) Comparison of large crater and multiringed basin populations on Mars, Mercury, and the Moon. *Proc. Lunar Sci. Conf.* **7**, 3589–3602.

Marvin U.B. (1976) Sample 14082 - Hand specimen descriptions and documentation. In *Interdisciplinary Studies by the Imbrium Consortium* **1**, Lunar Science Institute Contr. **267D**, 40–41.

Marvin U.B. and Lindstrom M.M. (1983) Rock 67015: A feldspathic fragmental breccia with KREEP-rich melt clasts. *Proc. Lunar Planet. Sci. Conf.* **13**, *J. Geophys. Res.* **88**, A659–A670.

Masaitis V.L., Mikhaylov M.V., and Selivanovskaya T.V. (1976) *The Popigay meteorite crater.* NASA Technical Translation **F–16900**, 167 pp.

Masaitis V.L., Danilin A.N., Mashchak M.S., Raikhlin A.I., Selivanovskaya T.V., and Shadenkov E.M. (1980) *Geologia Astroblem* (The Geology of Astroblemes), Nedra Press, Leningrad, 231 pp. (in Russian).

Mason R., Guest J.E., and Cooke G.N. (1976) An Imbrium pattern of graben on the Moon. *Proc. Geol. Assoc.* **87** (UK), 161–168.

Maurer P., Eberhardt P., Geiss J., Grogler N., Stettler A., Brown G., Peckett A., and Krahenbuhl U. (1978) Pre-Imbrian craters and basins: ages, compositions and excavation depths of Apollo 16 breccias. *Geochim. Cosmochim. Acta* **42**, 1687–1720.

Maxwell D.E. (1977) Simple Z model of cratering, ejection and overturned flap. In *Impact and Explosion Cratering* (D.J. Roddy, R.O. Pepin, and R.B. Merrill, editors), pp. 1003–1008, Pergamon Press, N.Y.

Maxwell T.A., El-Baz F., and Ward S.H. (1975) Distribution, morphology and origin of ridges and arches in Mare Serenitatis. *Bull. Geol. Soc. America* **86**, 1273–1278.

McCauley J.F. (1967) Geologic map of the Hevelius region of the Moon. *U.S. Geol. Survey Map* **I–491**.

McCauley J.F. (1968) Advance systems traverse research project report (G.E. Ulrich, editor) U.S. Geol. Survey Interagency Report *Astrogeology* **7**, 59 pp.

McCauley J.F. (1977) Orientale and Caloris. *Phys. Earth Planet. Interiors* **15**, 220–250.

McCauley J.F., Guest J.E., Schaber G.G., Trask N.J., and Greeley R. (1981) Stratigraphy of the Caloris basin, Mercury. *Icarus* **47**, 184–202.

McCord T.B., Charette M.P., Johnson T.V., Lebofsky L.A., Pieters C., and Adams, J.B. (1971) Lunar spectral types. *J. Geophys. Res.* **77**, 1349–1359.

McCord T.B., Clark R N., Hawke B.R., McFadden L.A., Owensby P.D., Pieters C.M., and Adams J.B. (1981) Near-infrared spectral reflectance, a first good look. *J. Geophys. Res.* **86**, 10883–10892.

McCormick K., Taylor G.J., Keil K., Spudis P.D., Grieve R.A.F., and Ryder G. (1989)

Sources of clasts in terrestrial impact melts: Clues to the origin of LKFM. *Proc. Lunar Planet. Sci. Conf.* **19**, 691–696.

McGetchin T.R., Settle M., and Head J.W. (1973) Radial thickness variation in impact crater ejecta: Implications for lunar basin deposits. *Earth Planet. Sci. Letters* **20**, 226–236.

McGill G.E. (1989) Buried topography of Utopia, Mars: Persistence of a giant impact depression. *J. Geophys. Res.* **94**, 2753–2759.

McKinley J.P., Taylor G.J., Keil K., Ma M.S., and Schmitt R.A. (1984) Apollo 16: Impact melt sheets, contrasting nature of the Cayley plains and Descartes mountains, and geologic history. *Proc. Lunar Planet. Sci. Conf.* **14**, *J. Geophys. Res.* **89**, B513–B524.

McKinnon W.B. (1981) Application of ring tectonic theory to Mercury and other solar system bodies. In *Multi-ring Basins* (P.H. Schultz and R.B. Merrill, editors), *Proc. Lunar Planet. Sci.* **12A**, 259–273.

McKinnon W.B. and Melosh H.J. (1980) Evolution of planetary lithospheres: Evidence from multiringed structures on Ganymede and Callisto. *Icarus* **44**, 454–471.

McSween H.Y. (1987) *Meteorites and their Parent Planets.* Cambridge University Press, Cambridge, 237 pp.

Melosh H.J. (1976) On the origin of fractures radial to lunar basins. *Proc. Lunar Sci. Conf.* **7**, 2967–2982.

Melosh H.J. (1979) Acoustic fluidization: A new geologic process? *J. Geophys. Res.* **84**, 7513–7520.

Melosh H.J. (1989) *Impact Cratering: A Geologic Process.* Oxford University Press, New York, 245 pp.

Melosh H.J. and McKinnon W.B. (1978) The mechanics of ringed basin formation. *Geophys. Res. Letters* **5**, 985–988.

Melosh H.J. and McKinnon W.B. (1988) Tectonics. In *Mercury* (C. Chapman and F. Vilas, editors), Univ. of Arizona Press, Tucson, 374–400.

Metzger A.E., Trombka J.I., Peterson L.E., Reedy R.C., and Arnold J.R. (1973) Lunar surface radioactivity: Preliminary results of the Apollo 15 and gamma-ray spectrometer experiments. *Science* **179**, 800 803.

Metzger A.E., Haines E.L., Parker R.E., and Radocinski R.G. (1977) Thorium concentrations in the lunar surface. I: Regional values and crustal content. *Proc. Lunar Sci. Conf.* **8**, 949–999.

Metzger A.E., Haines E.L., Etchegaray-Ramirez M.I., and Hawke B.R. (1979) Thorium concentrations in the lunar surface: III. Deconvolution of the Apenninus region. *Proc. Lunar Planet. Sci. Conf.* **10**, 1701–1718.

Metzger A.E., Etchegaray-Ramirez M.I., and Haines E.L. (1981) Thorium concentration in the lunar surface: V. Deconvolution of the central highlands region. *Proc. Lunar Planet. Sci.* **12B**, 751–766.

Meyer C. (1977) Petrology, mineralogy and chemistry of KREEP basalt. *Phys. Chem. Earth* **10**, 239–260.

Milton D.J. (1972) Geologic maps of the Descartes region of the Moon (Apollo 16 pre-mission maps). *U.S. Geol. Survey Map* I–**748** (sheet 1).

Moore H.J. (1967) Geologic map of the Seleucus quadrangle of the Moon. *U.S. Geol. Survey Map* I–**527**.

Moore H.J., Hodges C.A., and Scott D.H. (1974) Multi-ringed basins -illustrated by Orientale and associated features. *Proc. Lunar Sci. Conf.* **5**, 71–100.

Moore J.M. and Horner V.M. (1984) The geomorphology of Rhea. In *Lunar and Planetary Science* XV, pp. 560–561, The Lunar and Planetary Institute, Houston.

Moore J.M., Spudis P.D., Pike R.J., and Greeley R. (1984) Multi-ringed basins of the saturnian satellites. *Geol. Soc. America Abs. Program.* **16**, 600.

Moore J.M., Horner V.M., and Greeley R. (1985) The geomorphology of Rhea: Implications for geologic history and surface processes. *Proc. Lunar Planet. Sci. Conf.* **15**, J. *Geophys. Res.* **90**, C785–C795.

Morgan J.W., Ganapathy R., Higuchi H., Krahenbuhl U., and Anders E. (1974) Lunar basins: Tentative characterization of projectiles, from meteoritic elements in Apollo 17 boulders. *Proc. Lunar Sci. Conf.* **5**, 1703–1736.

Morris E.C. and Wilhelms D.E. (1967) Geologic map of the Julius Caesar quadrangle of the Moon. *U.S. Geol. Survey Map* **I–510**.

Morrison R.H. and Oberbeck V.R. (1975) Geomorphology of crater and basin deposits - emplacement of the Fra Mauro formation. *Proc. Lunar Sci. Conf.* **6**, 2503–2530.

Murali A.V., Ma M.S., Laul J.C., and Schmitt R.A. (1977) Chemical composition of breccias, feldspathic basalt and anorthosites from Apollo 15 (15308, 15359, 15382, and 15362), Apollo 16 (60618 and 65785), Apollo 17 (72435, 72536, 72559, 72735, 72738, 78526, and 78527) and Luna 20 (22012 and 22013). In *Lunar Science* **VIII**, pp. 700–702. Lunar Science Institute, Houston.

Murray B.C., Belton M.J.S., Danielson G.E., Davies M.E., Gault D.E., Hapke B., O'Leary B., Strom R.G., Suomi V., and Trask N. (1974) Mercury's surface: Preliminary description and interpretation from Mariner 10 pictures. *Science* **185**, 169–179.

Murray J.B. (1980) Oscillating peak model of basin and crater formation. *Moon and Planets* **22**, 269–291.

Mutch T.A., Arvidson R.E., Head J.W., Jones K.L., and Saunders R.S (1976) *The Geology of Mars.* Princeton University Press, Princeton N.J., 400 pp.

Norman M.D. and Ryder G. (1979) A summary of the petrology and geochemistry of pristine highlands rocks. *Proc. Lunar Planet Sci. Conf.* **10**, 531–559.

Nyquist L.E., Bansal B.M, and Wiesmann H. (1975) Rb–Sr ages and initial $^{87}Sr/^{86}Sr$ for Apollo 17 basalts and KREEP basalt 15386. *Proc. Lunar Sci. Conf.* **6**, 1445–1465.

Oberbeck V.R. (1975) The role of ballistic erosion and sedimentation in lunar stratigraphy. *Rev. Geophys. Space Phys.* **13**, 337–362.

Orphal D.L. and Schultz P.H. (1978) An alternative model for the Manicouagan impact structure. *Proc. Lunar Planet. Sci. Conf.* **9**, 2695–2712.

Page N.J. (1970) Geologic map of the Cassini quadrangle of the Moon. *U.S. Geol. Survey Map* **I–666**.

Papike J.J., Simon S.B., and Laul J.C. (1982) The lunar regolith: Chemistry, mineralogy, and petrology. *Rev. Geophys. Space Phys.* **20**, 761–826.

Peterson J.E. (1977) Geologic map of the Noachis quadrangle of Mars. *U.S. Geol. Survey Map* **I–910**.

Peterson J.E. (1978) Volcanism in the Noachis - Hellas region of Mars, 2. *Proc. Lunar Planet. Sci. Conf.* **9**, 3311–3432.

Phillips R.J., Arvidson R.E., Boyce J.M., Campbell D.B., Guest J.E., Schaber G.G., and Soderblom L.A. (1991) Impact craters on Venus: Initial analysis from Magellan. *Science* **252**, 288–297.

Phillips R.J., Raubertas R.F., Arvidson R.E., Sarkar I.C., Herrick R.R., Izenberg N., and

Grimm R.E. (1992) Impact craters and Venus resurfacing history. *J. Geophys. Res.* **97** E10, 15923–15948.

Phinney W.C, Dence M.R., and Grieve R.A.F. (1978) Investigation of the Manicouagan crater, Quebec: An introduction. *J. Geophys. Res.* **83**, 2729–2735.

Pieters C.M. (1986) Composition of the lunar highlands crust from near-infrared spectroscopy. *Rev. Geophysics* **24**, 557–578.

Pike R.J (1971) Genetic implications of the shapes of martian and lunar craters. *Icarus* **15**, 384–395.

Pike R.J. (1974) Ejecta from large craters on the Moon: Comments on the geometric model of McGetchin *et al. Earth Planet. Sci. Letters* **23**, 265–274.

Pike R.J. (1980a) Geometric interpretation of lunar craters. *U.S. Geol. Survey Prof. Paper* **1046**-C, 77 pp.

Pike R.J. (1980b) Formation of complex impact craters: Evidence from Mars and other planets. *Icarus* **43**, 1–19.

Pike R.J. (1980c) Control of crater morphology by gravity and target type: Mars, Earth, Moon. *Proc. Lunar Planet. Sci. Conf.* **11**, 2159–2189.

Pike R.J. (1982) Crater peaks to basin rings: The transition on Mercury and other bodies. In *Reports of Planetary Geology Program –1982*, pp. 117–119, NASA **TM-85127**.

Pike R.J. (1983) Large craters or small basins on the Moon (abstract) In *Lunar and Planetary Science* **XIV**, p. 610–611, Lunar and Planetary Institute, Houston.

Pike R.J. (1985) Some morphologic systematics of complex impact structures. *Meteoritics* **20**, 49–68.

Pike R.J. (1988) Geomorphology of craters on Mercury. In *Mercury* (C. Chapman and F. Vilas, editors), Univ. of Arizona Press, Tucson, 165–273.

Pike R.J. and Spudis P.D. (1987) Basin-ring spacing on the Moon, Mercury, and Mars. *Earth Moon and Planets* **39**, 129–194.

Podosek F.A., Huneke, J.C., Gancarz A.J., and Wasserburg G.J. (1973) The age and petrography of two Luna 20 fragments and inferences for widespread lunar metamorphism. *Geochim. Cosmochim. Acta* 37, 887–904.

Pohl J., Stöffler D., Gall H., and Ernstson K. (1977) The Ries impact crater. In *Impact and Explosion Cratering* (D.J. Roddy, R.O. Pepin, and R.B. Merrill, editors), pp. 343–404, Pergamon Press, N.Y.

Potter D.B. (1976) Geologic map of the Hellas quadrangle of Mars. *U.S. Geol. Survey Map* **I-941**.

Prinz M., Dowty E., Keil K., and Bunch T.E. (1973) Mineralogy, petrology and chemistry of lithic fragments from Luna 20 fines: Origin of the cumulate ANT suite and its relationship to high-alumina and mare basalts. *Geochim. Cosmochim. Acta* **37**, 979–1006.

Proctor R.A. (1873) *The Moon: Her motions, aspect, scenery, and physical condition.* Alfred Brothers, Manchester, U.K., 314 pp.

Quaide W.L. and Oberbeck V.R. (1968) Thickness determinations of the lunar surface layer from lunar impact craters. *J. Geophys. Res.* **73**, 5247–5270.

Reed V.S. (1981) Geology of areas near South Ray and Baby Ray craters. *In* Geology of the Apollo 16 Area, Central lunar highlands. *U.S. Geol. Survey Prof. Paper* **1048**, pp. 82–105.

Reed V.S. and Wolfe E.W. (1975) Origin of the Taurus-Littrow massifs. *Proc. Lunar Sci. Conf.* **6**, 2443–2461.

Reid A.M., Warner J., Ridley W.I., and Brown R.W. (1972) Major element composition of glasses in three Apollo 15 soils. *Meteoritics* **7**, 395–415.

Reid A.M., Duncan A.R., and Richardson S.H. (1977) In search of LKFM. *Proc. Lunar Sci. Conf.* **8**, 2321–2338.

Ridley W.I. (1976) Petrology of lunar rocks and implication to lunar evolution. *Ann. Rev. Earth Planet. Sci.* **4**, 15–48.

Ridley W.I., Hubbard N.J., Rhodes J.M., Wiesmann H., and Bansal B. (1973) The petrology of lunar breccia 15445 and petrogenetic implications. *J. Geology* **81**, 621–631.

Robertson P.B. and Sweeney J.F. (1983) Haughton impact structure: structural and morphologic aspects. *Canadian Jour. Earth Sci.* **20**, 1134–1151.

Roddy D.J. (1977) Large-scale impact and explosion craters: Comparisons of morphological and structural analogs. In *Impact and Explosion Cratering* (R.J. Roddy, R.O. Pepin, and R.B. Merrill, editors), pp.185–246, Pergamon Press, N.Y.

Rükl, A. (1972) *Maps of lunar hemispheres*. D. Reidel, Dordrecht, 24 pp., 6 plates.

Runcorn S.K. (1982) Primeval displacements of the lunar pole. *Phys. Earth Planet. Interiors* **29**, 135–147.

Ryder G. (1976) Lunar sample 15405: Remnant of a KREEP basalt-granite differentiated pluton. *Earth Planet Sci. Letters* **29**, 255–268.

Ryder G. (1981a) Apollo 16 basaltic impact melts: Chemistry and relationships. In *Workshop on Apollo 16* (O.B. James and F. Hörz, editors), pp. 108–111. LPI Tech. Rpt. **81–01**, Lunar and Planetary Institute, Houston.

Ryder G. (1981b) The Apollo 17 highlands: The South Massif soils. In *Lunar and Planetary Science* **XII**, pp. 918–920. Lunar and Planetary Institute, Houston.

Ryder G. (1987) Petrographic evidence for nonlinear cooling rates and a volcanic origin for Apollo 15 KREEP basalts. *Proc. Lunar Planet. Sci. Conf.* **17**, *J. Geophys. Res.* **92**, E331–E339.

Ryder G. (1988) Quenching and disruption of lunar KREEP lava flows by impacts. *Nature* **336**, 751–754.

Ryder G. (1990) Lunar samples, lunar accretion, and the early bombardment of the Moon. *EOS, Trans. Amer. Geophys. Union* **71**, 313–333.

Ryder G. and Bower J.F. (1976) Sample 14082 – Petrology. In *Interdisciplinary Studies by the Imbrium Consortium* **1**, Lunar Science Institute Contr. **267D**, 41–50.

Ryder G. and Bower J.F. (1977) Petrology of Apollo 15 black-and-white rocks 15445 and 15455–Fragments of the Imbrium impact welt sheet? *Proc. Lunar Sci. Conf.* **8**, 1895–1923.

Ryder G. and Norman M. (1978a) *Catalog of pristine non-mare materials. Part 1. Non-anorthosites.* NASA **JSC–14565**, 146 pp.

Ryder G. and Norman M. (1978b) *Catalog of pristine non-mare materials. Part 2. Anorthosites.* NASA **JSC–14603**, 86 pp.

Ryder G. and Norman M.D. (1980) *Catalog of Apollo 16 rocks.* NASA **JSC–16904**, 1144 pp.

Ryder G. and Spudis P. (1980) Volcanic rocks in the lunar highlands. *Proc. Conf. Lunar Highlands Crust* (J.J. Papike and R.B. Merrill, editors), pp. 353–375, Pergamon Press, N.Y.

Ryder G. and Spudis P.D. (1987) Chemical composition and origin of Apollo 15 impact melts. *Proc. Lunar Planet. Sci. Conf.* **17**, *J. Geophys. Res.* **92**, E432–E446.

Ryder G. and Taylor G.J. (1976) Did mare volcanism commence early in lunar history? *Proc. Lunar Sci. Conf.* **7**, 1741–1755.

Ryder G. and Wood J.A. (1977) Serenitatis and Imbrium impact melts: Implications for large-scale layering in the lunar crust. *Proc. Lunar Sci. Conf.* **8**, 655–668.

Ryder G., Stoeser D.B., Marvin U.B., Bower J.F., and Wood J.A. (1975) Boulder 1, Station 2, Apollo 17: Petrology and petrogenesis. *The Moon* **14**, 327–357.

Ryder G., Stoeser D.B., and Wood J.A. (1977) Apollo 17 KREEP basalt: A rock type intermediate mare and KREEP basalt. *Earth Planet. Sci. Letters* **35**, 1–13.

Saunders R.S., Arvidson R.E., Head J.W., Schaber G.G., Stofan E.R., and Solomon S.C. (1991) An overview of Venus geology. *Science* **252**, 249–252.

Schaber G.G., Boyce J.M., and Trask N.J. (1977) Moon–Mercury: Large impact structures, isostasy, and average crustal viscosity. *Phys. Earth Planet. Interiors* **15**, 189–201.

Schaber G.G., Shoemaker E.M., and Kozak R.C. (1987) The surface age of Venus: Use of the terrestrial cratering record. *Astronomicheskii Vestnik* **21**, 144–151.

Schaber G.G., Strom R.G., Moore H.J., Soderblom L.A., Kirk R.L., Chadwick D.J., Dawson D.D., Gaddis L.R., Boyce J.M., and Russell J. (1992) Geology and distribution of impact craters on Venus: What are they telling us? *J. Geophys. Res.* **97**, E8, 13257–13301.

Schmidt-Kaler H., Treibs W., and Huttner R. (1970) *Geologische Übersichtskarte des Rieses und seiner Umgebung* 1:100,000. Exkursions-führer zur Übersichtskarte des Rieses 1:100,000. Bayerisches Geologisches Landesamt, München, 68 pp.

Schmitt H.H. (1973) Apollo 17 report on the valley of Taurus-Littrow. *Science* **182**, 681–690.

Schonfeld E. (1974) The contamination of lunar highland rocks by KREEP: Interpretation by mixing models. *Proc. Lunar Sci. Conf.* **5**, 1269–1286.

Schonfeld E. (1977) Comparison of orbital chemistry with crustal thickness and lunar sample chemistry. *Proc. Lunar Sci. Conf.* **8**, 1149–1162.

Schonfeld E. (1980) Enhanced orbital geochemical images by the Laplacian subtraction method. *Proc. Lunar Planet. Sci. Conf.* **11**, 677–688.

Schonfeld E. (1981) High spatial resolution Mg/Al maps of the western Crisium and Sulpicius Gallus regions. *Proc. Lunar Planet. Sci.* **12B**, 809–816.

Schonfeld E. and Meyer C. (1972) The abundances of components of the lunar soils by a least-squares mixing model and the formation age of KREEP. *Proc. Lunar Sci. Conf.* **3**, 1397–1420.

Schonfeld E. and Meyer C. (1973) The old Imbrium hypothesis. *Proc. Lunar Sci. Conf.* **4**, 125–138.

Schultz P.H. (1976a) *Moon Morphology*. Univ. of Texas Press, Austin, 626 pp.

Schultz P.H. (1976b) Floor-fractured lunar craters. *The Moon* **15**, 241–273.

Schultz P.H. (1978) Ejecta dynamics of large-scale impacts. In *Lunar and Planetary Science IX*, pp. 1024–1026. Lunar and Planetary Institute, Houston.

Schultz P.H. (1979) Evolution of intermediate-age impact basins on the Moon. In *Papers Presented to the Conference on the Lunar Highlands Crust*, pp. 141–142, Lunar and Planetary Institute, Houston.

Schultz P.H. (1984) Impact basin control of volcanic and tectonic features on Mars. In *Lunar and Planetary Science* **XV**, pp. 728–729, Lunar and Planetary Institute, Houston.

Schultz P.H. and Gault D.E. (1975) Seismic effects from major basin formation on the Moon and Mercury. *The Moon* **12**, 159–177.

Schultz P.H. and Gault D.E (1985) Clustered impacts: Experiments and implications. *J. Geophys. Res.* **90**, 3701–3732.

Schultz P.H. and Gault D.E. (1986) Experimental evidence for non-proportional growth of large craters. In *Lunar and Planetary Science* **XVII**, pp. 777–778, Lunar and Planetary Institute, Houston.

Schultz P.H. and Glicken H. (1979) Impact crater and basin control of igneous processes on Mars. *J. Geophys. Res.* **84**, 8033–8047.

Schultz P.H. and Mendell W. (1978) Orbital infrared observations of lunar craters and possible implications for impact ejecta emplacement. *Proc. Lunar Planet. Sci. Conf.* **9**, 2857–2883.

Schultz P.H. and Mendenhall M.H. (1979) On the formation of basin secondary craters by ejecta complexes. In *Lunar and Planetary Science* **X**, pp. 1078–1080. Lunar and Planetary Institute, Houston.

Schultz P.H. and Spudis P.D. (1978) The dark ring of Orientale: Implications for pre-basin mare volcanism and a clue to the identification of the transient cavity rim. In *Lunar and Planetary Science* **IX**, pp. 1033–1035. Lunar and Planetary Institute, Houston.

Schultz P.H. and Spudis P.D. (1979) Evidence for ancient mare volcanism. *Proc. Lunar Planet. Sci. Conf.* **10**, 2899–2918.

Schultz P.H. and Spudis P.D. (1983) The beginning and end of lunar mare volcanism. *Nature* **302**, 233–236.

Schultz P.H. and Spudis P.D. (1985) Procellarum basin: A major impact or the effect of Imbrium? In *Lunar and Planetary Science* **XVI**, pp. 746–747, Lunar and Planetary Institute, Houston.

Schultz P.H. and Srnka L.J. (1980) Cometary collisions on the Moon and Mercury. *Nature* **284**, 22–26.

Schultz P.H., Orphal D., Miller B., Borden W.F., and Larson S.A. (1981) Multi-ring basin formation: Possible clues from impact cratering calculations. In *Multi-ring Basins* (P.H. Schultz and R.B. Merrill, editors), *Proc. Lunar Planet Sci.* **12A**, 181–195.

Schultz P.H., Schultz R.A., and Rogers J. (1982) Structure and evolution of ancient impact basins on Mars. *J. Geophys. Res.* **87**, 9803–9820.

Schultz P.H., Gault D.E., and Crawford D. (1986) Impacts of hemispherical granular targets: Implications for global impacts. In *Lunar and Planetary Science* **XVII**, pp. 783–784, Lunar and Planetary Institute, Houston.

Schultz R.A. and Frey H.V. (1990) A new survey of multiring impact basins on Mars. *J. Geophys. Res.* **95**, 14175–14189.

Scott D.H. (1972a) Structural aspects of Imbrium sculpture. In *Apollo 16 Prelim. Sci. Rpt.*, NASA **SP–315**, pp. 29–31 to 29–33.

Scott D.H. (1972b) Geologic map of the Eudoxus quadrangle of the Moon. *U.S. Geol. Survey Map* **I–705**.

Scott D.H. (1974) The geologic significance of some lunar gravity anomalies. *Proc. Lunar Sci. Conf.* **5**, 3025–3036.

Scott D.H. and Carr M.H. (1972) Geologic maps of the Taurus-Littrow region of the Moon (Apollo 17 pre-mission maps). *U.S. Geol. Survey Map* **I–800** (sheet 1).

Scott D.H. and Carr M.H. (1978) Geologic map of Mars. *U.S. Geol. Survey Map* **I–1083**.

Scott D.H. and Tanaka K.L. (1981) A large highland volcanic province revealed by Viking images. *Proc. Lunar Planet. Sci.* **12B**, 1449–1458.

Scott D.H., McCauley J.F., and West M.N. (1977) Geologic map of the west side of the Moon. *U.S. Geol. Survey Map* **I–1034**.

Sharpton V.I and Ward P.D., editors (1990) *Global catastrophes in Earth history: An interdisciplinary conference on impacts, volcanism, and mass mortality.* Geol. Soc. America Special Paper **247**, 631 pp.

Sharpton V.I., Dalrymple G.B., Ryder G., Schuraytz B.C., and Urrtia-Fucugauchi J. (1992) The Chicxulub impact structure and the Cretaceous-Tertiary boundary. *Nature* **359**, 819–821.

Shoemaker E.M. (1962) Interpretation of lunar craters. In *Physics and Astronomy of the Moon* (Z. Kopal, editor), pp. 283–359, Academic Press, N.Y.

Shoemaker E.M. (1972) Cratering history and early evolution of the Moon. In *Lunar Science III*, pp. 696–698. Lunar Science Institute, Houston.

Shoemaker E.M. (1977) Why study impact craters? In *Impact and Explosion Cratering* (D.J. Roddy, R.O. Pepin, and R.B. Merrill, editors), pp. 1–10, Pergamon Press, N.Y.

Shoemaker E.M. and Chao E.C.T. (1961) New evidence for the impact origin of the Ries Basin, Bavaria, Germany. *J. Geophys. Res.* **66**, 3371–3378.

Shoemaker E.M. and Hackman R.J. (1962) Stratigraphic basis for a lunar time scale. In *The Moon* (Z. Kopal and Z.K. Mikhailov, editors), pp. 289–300, Academic Press, N.Y.

Shoemaker E.M. and Shoemaker C.S. (1990) The collision of solid bodies. In *The New Solar System* (J.K. Beatty and A. Chaikin, editors), Sky Publishing and Cambridge Univ. Press, 259–274.

Shoemaker E.M., Batson R.M., Holt H.E., Morris E.C., Rennilson J.J., and Whitaker, E.A. (1968) Television observations from Surveyor VII. In *Surveyor VII Mission Report*, Part II. Science Results. Jet Propul. Lab. Tech. Rept. **32–1264**, 9–76.

Shoemaker E.M., Lucchitta B.K., Wilhelms D.E., Plescia J.B. and Squyres S. (1982) The geology of Ganymede. In *Satellites of Jupiter* (D. Morrison, editor), Univ. of Arizona, Tucson, pp. 435–520.

Short N.M. and Forman M.L. (1972) Thickness of impact crater ejecta on the lunar surface. *Modern Geology* **3**, 69–91.

Silver L.T. and Schultz P.H., editors (1982) *Geological implications of impacts of large asteroids and comets on the Earth.* Geol Soc. America Special Paper **190**, 528 pp.

Simonds C.H. (1975) Thermal regimes in impact melts and the petrology of the Apollo 17 Station 6 boulder. *Proc. Lunar Sci. Conf.* **6**, 641–672.

Simonds C.H., Warner J.L., and Phinney W.C. (1973) Petrology of Apollo 16 poikilitic rocks. *Proc. Lunar Sci. Conf.* **4**, 613–632.

Simonds C.H., Phinney W.C., and Warner J.L. (1974) Petrography and classification of Apollo 17 non-mare rocks with emphasis on samples from the Station 6 boulder. *Proc. Lunar Sci. Conf.* **5**, 337–353.

Simonds C.H., Warner J.L., Phinney W.C., McGee P.E. (1976) Thermal model for impact breccia lithification: Manicouagan and the Moon. *Proc. Lunar Sci. Conf.* **7**, 2509–2528.

Simonds C.H., Phinney W.C., Warner J.L., McGee P.E., Geeslin J., Brown R.W., and Rhodes J.M. (1977) Apollo 14 revisited, or breccias aren't so bad after all. *Proc. Lunar Sci. Conf.* **8**, 1869–1893.

Simonds C.H., Floran R.J., McGee P.E., Phinney W.C., and Warner J.L. (1978) Petrogenesis of melt rocks, Manicouagan impact structure, Quebec. *J. Geophys. Res.* **83**, 2773–2788.

Sjogren W.L., Wimberly R.N., and Wollenhaupt W.R. (1974) Lunar gravity via the Apollo 15 and 16 subsatellites. *The Moon* **9**, 115–128.

Smith B.A. and 21 others (1979) The Jupiter system through the eyes of Voyager 1. *Science* **204**, 13–32.

Smith B.A. and 25 others (1981) Encounter with Saturn: Voyager 1 imaging science results. *Science* **212**, 163–191.

Smith M.R., Schmitt R.A., Warren P.H., Taylor G.J., and Keil K. (1983) Far-eastern non-mare samples: New data from Luna 20 and 16. *Lunar Planetary Science* XIV, 716–717.

Solomon S.C. and Head J.W. (1980) Lunar mascon basins: Lava filling, tectonics and evolution of the lithosphere. *Rev. Geophys. Space Phys.* **18**, 107–141.

Solomon S.C., Comer R.P., and Head J.W. (1982) The evolution of impact basins: Viscous relaxation of topographic relief. *J. Geophys. Res.* **87**, 3975–3992.

Spudis P.D. (1978) Composition and origin of the Apennine Bench Formation. *Proc. Lunar Planet. Sci. Conf.* **9**, 3379–3394.

Spudis P.D. (1979) The extent and duration of lunar KREEP volcanism. In *Papers Presented to the Conference on the Lunar Highlands Crust*, pp. 157–159. Lunar and Planetary Institute, Houston.

Spudis P.D. (1980) Petrology of the Apennine Front, Apollo 15: Implications for the geology of the Imbrium basin. In *Papers Presented to the Conference on Multi-ring Basins: Formation and Evolution*, pp. 83–85. Lunar and Planetary Institute, Houston.

Spudis P.D. (1981) The nature of lunar basin ejecta deposits inferred from Apollo highland landing site geology. In *Reports of Planetary Geology Program-1981*, pp. 120–122. NASA **TM–84211**.

Spudis P.D. (1982) Orientale basin ejecta: Depths of derivation and implications for the basin-forming process. In *Lunar and Planetary Science* **XIII**, pp. 760–761. Lunar and Planetary Institute, Houston.

Spudis P.D. (1983) Mercury: New identification of ancient multi-ring basins and implications for geologic evolution. In *Reports of Planetary Geology Program*, pp. 87–89, NASA **TM-86246**.

Spudis P.D. (1984) Apollo 16 site geology and impact melts: Implications for the geologic history of the lunar highlands. *Proc. Lunar Planet. Sci.* **15**, *J. Geophys. Res.* **89**, C95–C107.

Spudis P.D. (1986) Materials and formation of the Imbrium basin. In *Workshop on the Geology and Petrology of the Apollo 15 Landing Site* (P.D. Spudis and G. Ryder, editors), pp. 100–104, LPI Tech. Rpt. **86-03**, Lunar and Planetary Institute, Houston.

Spudis P.D. and Davis P.A. (1983) Identification of regional deposits of lunar pristine rocks from orbital geochemical data. *Workshop on Pristine Highlands Rocks* (J. Longhi and G. Ryder, editors), Lunar and Planetary Institute, Tech. Report **83-02**, 69–71.

Spudis P.D. and Davis P.A. (1985) How much anorthosite is in the lunar crust?: Implications for lunar crustal origin. In *Lunar and Planetary Science* **XVI**, pp. 807–808, Lunar and Planetary Institute, Houston.

Spudis P.D. and Davis P.A. (1986) A chemical and petrological model of the lunar crust and implications for lunar crustal origin. *Proc. Lunar Planet. Sci. Conf.* **17**, *J. Geophys. Res.* **91**, E84–E90.

Spudis P.D. and Guest J.E. (1988) Stratigraphy and geologic history of Mercury. In *Mercury*, (F. Vilas, C. Chapman and M. Matthews, editors), Univ. of Arizona Press, Tucson, pp. 118–164 and 710–715 (color plates).

Spudis P.D. and Hawke B.R. (1981) Chemical mixing model studies of lunar orbital geochemical data: Apollo 16 and 17 highlands compositions. *Proc. Lunar Planet. Sci.* **12B**, 781–789.

Spudis P.D. and Hawke B.R. (1986) The Apennine Bench Formation revisited. In *Workshop on the Geology and Petrology of the Apollo 15 Landing Site* (P.D. Spudis and G. Ryder, editors), pp.105–107, LPI Tech. Rpt. **86-03**, Lunar and Planetary Institute, Houston.

Spudis P.D. and Head J.W. (1977) Geology of the Imbrium basin Apennine mountains and relation to the Apollo 15 landing site. *Proc. Lunar Sci. Conf.* **8**, 2785–2797.

Spudis P.D. and Ryder G. (1981) Apollo 17 impact melts and their relation to the Serenitatis basin. In *Multi-ring Basins* (P.H. Schultz and R.B. Merrill, editors), *Proc. Lunar Planet. Sci.* **12A**, 133–148.

Spudis P.D. and Ryder G. (1985) Geology and petrology of the Apollo 15 landing site: Past, present, and future understanding. *EOS Trans. Amer. Geophys. Union* **66**, 721–726.

Spudis P.D. and Ryder G., editors (1986) *Workshop on the geology and petrology of the Apollo 15 landing site.* LPI Tech. Rpt. **86-03**, Lunar and Planetary Institute, Houston, 126 pp.

Spudis P.D. and Schultz P.H. (1985) The proposed lunar Procellarum basin: Some geochemical inconsistencies. In *Lunar and Planetary Science* **XVI**, pp. 809–810, Lunar and Planetary Institute, Houston.

Spudis P.D., Cintala M.J., and Grieve R.A.F. (1984a) The early Moon: Implications of a large impact into a hot target. In *Lunar and Planetary Science* **XV**, pp. 810–811, Lunar and Planetary Institute, Houston.

Spudis P.D., Hawke B.R., and Lucey P.G. (1984b) Composition of Orientale basin deposits and implications for the lunar basin-forming process. *Proc. Lunar Planet. Sci. Conf.* **15**, *J. Geophys. Res.* **89**, C197–C210.

Spudis P.D., Davis P.A., and Pattanaborwornsak B. (1988a) The Fe–Al relation of lunar soils and orbital chemical data: Implications for Al abundances estimated from Apollo orbital gamma-ray data. In *Lunar and Planetary Science* **XIX**, pp. 1113–1114, Lunar and Planetary Institute, Houston.

Spudis P.D., Hawke B.R., and Lucey P.G. (1988b) Materials and formation of the Imbrium basin. *Proc. Lunar Planet. Sci. Conf.* **18**, Lunar and Planetary Institute and Cambridge Univ. Press, 155–168.

Spudis P.D., Hawke B.R., and Lucey P.G. (1989) Geology and deposits of the lunar Nectaris basin. *Proc. Lunar Planet. Sci. Conf.* **19**, 51–60.

Spudis P.D., Ryder G., Taylor G.J., McCormick K.A., Keil K., and Grieve R.A.F. (1991) Sources of mineral fragments in impact melts 15445 and 15455: Toward the origin of low-K Fra Mauro basalt. *Proc. Lunar Planet. Sci. Conf.* **21**, 151–165.

Spudis P.D., Hawke B.R., Lucey P.G., Taylor G.J., and Peterson C. (1992) Geology and deposits of the Humorum basin. *Lunar and Planetary Science* **XXIII**, 1345–1346.

Stadermann F.J., Heusser E., Jessberger E.K., Lingner S., and Stöffler D. (1991) The case for a younger Imbrium basin: New ^{40}Ar–^{39}Ar ages of Apollo 14 rocks. *Geochim. Cosmochim. Acta* **55**, 2339–2349.

Stam M., Schultz P.H., and McGill G.E. (1984) Martian impact basins: Morphology differences and tectonic provinces. *Lunar Planetary Science* **XV**, 818–819.

Stöffler D., Gault D.E., Wedekind J., and Polkowski G. (1975) Experimental hypervelocity impact into quartz sand: Distribution and shock metamorphism of ejecta. *J. Geophys. Res.* **80**, 4062–4077.

Stöffler D., Bischoff A., Borchadt R., Burghele A., Deutsch A., Jessberger E.K., Ostertag R.,

Palme H., Spettel B., Reimold W.U., Wacker K., and Wänke H. (1985) Composition and evolution of the lunar crust in the Descartes highlands, Apollo 16. *Proc. Lunar Planet. Sci.* **15**, *J. Geophys. Res.* **90**, C449–C506.

Strain P.L. and El-Baz F. (1980) The geology and morphology of Ina. *Proc. Lunar Planet. Sci. Conf.* **11**, 2437–2446.

Strom R.G. (1984) Mercury. In *The Geology of the Terrestrial Planets* (M.H. Carr, editor), NASA **SP-469**, 12–55.

Strom R.G., Trask N.J., and Guest J.E. (1975) Tectonism and volcanism on Mercury. *J. Geophys. Res.* **80**, 2478–2507.

Stuart-Alexander D.E. (1971) Geologic map of the Reita quadrangle of the Moon. *U.S. Geol. Survey Map* **I–694**.

Stuart-Alexander D.E. (1978) Geologic map of the central far side of the Moon. *U.S. Geol. Survey Map* **I–1047**.

Stuart-Alexander D.E. and Howard K.A. (1970) Lunar maria and circular basins-a review. *Icarus* **12**, 440–456.

Stuart-Alexander D.E. and Wilhelms D.E. (1975) The Nectarian system, a new lunar time-stratigraphic unit. *J. Res. U.S. Geol. Survey* **3**, 53–58.

Swann G.A., Bailey N.G., Batson R.M., Eggleton R.E., Hait M.H., Holt H.E., Larson K.B., Reed V.S., Schaber G.G., Sutton R.L., Trask N.J., Ulrich G.E., and Wilshire H.G. (1977) Geology of the Apollo 14 landing site in the Fra Mauro highlands. *U.S. Geol. Survey Prof. Paper* **880**, 103 pp.

Swindle T., Spudis P.D., Taylor G.J., Korotev R., Nichols R.H., and Olinger C.T. (1991) Searching for Crisium basin ejecta: Chemistry and ages of Luna 20 impact melts. *Proc. Lunar Planet. Sci.* **21**, 167–184.

Swisher C.C., Grajales-Nishimura J.M., Montanari A., Margolis S.V., Claeys P., Alvarez W., Renne P., Cedillio-Pardo E., Maurrasse F., Curtis G.H., Smit J., and McWilliams M.O. (1992) Coeval $^{40}Ar/^{39}Ar$ ages of 65.0 million years ago from Chicxulub crater melt rock and Cretaceous-Tertiary boundary tektites. *Science* **257**, 954–958.

Taylor G.J., Drake M.J., Hallam M.E., Marvin U.B., and Wood J.A. (1973a) Apollo 16 stratigraphy: The ANT hills, the Cayley plains, and a pre-Imbrian regolith. *Proc. Lunar Sci. Conf.* **4**, 583–568.

Taylor G.J., Drake M.J., Wood J.A., and Marvin U.B. (1973b) The Luna 20 lithic fragments, and the composition and origin of the lunar highlands. *Geochim. Cosmochim. Acta* **37**, 1087–1106.

Taylor S.R. (1975) *Lunar Science: A Post Apollo View*. Pergamon Press, N.Y. 372 pp.

Taylor S.R. (1982) *Planetary Science: A Lunar Perspective*. Lunar and Planetary Institute Press, Houston, 481 pp.

Taylor S.R., Gorton M.P., Muir P., Nance W., Rudowski R., and Ware N. (1973) Lunar highlands composition: Apennine front. *Proc. Lunar Sci. Conf.* **4**, 1445–1459.

Tera F., Papanastassiou D., and Wasserberg G.J. (1974) Isotopic evidence for a terminal lunar cataclysm. *Earth Planet. Science Letters* **22**, 1–21.

Trask N.J. and Guest J.E. (1975) Preliminary geologic terrain map of Mercury. *J. Geophys. Res.* **80**, 2461–2477.

Trask N.J. and McCauley J.F. (1972) Differentiation and volcanism in the lunar highlands – Photogeologic evidence and Apollo 16 implications. *Earth Planet. Sci. Letters* **14**, 201–206.

Turner G. (1977) Potassium–Argon chronology of the Moon. *Phys. Chem. Earth* **10**, 145–195.

Ulrich G.E. and Reed V.S. (1981) Stratigraphic interpretations at the Apollo 16 site. In Geology of the Apollo 16 Area, Central lunar highlands. *U.S. Geol. Survey Prof. Paper* **1048**, pp. 197–214.

Urey H.C. (1952) *The Planets.* Yale University Press, New Haven, 245 pp.

Van Dorn W.G. (1968) Tsunamis on the Moon? *Nature* **220**, 1102–1107.

Warner J.L. (1972) Metamorphism of Apollo 14 breccias. *Proc. Lunar Sci. Conf.* **3**, 623–644.

Warner J.L., Reid A.M., Ridley W.I., and Brown R.W. (1972a) Major element composition of Luna 20 glasses. *Earth Planet. Sci. Letters* **17**, 7–12.

Warner J.L., Ridley W.I., Reid A.M., and Brown R.W. (1972b) Apollo 15 glasses and distribution of non-mare crustal rock types. In *The Apollo 15 Lunar Samples*, pp. 179–181. Lunar Science Institute, Houston.

Warner J.L., Simonds C.H., and Phinney W.C. (1973a) Apollo 16 rocks: Classification and petrogenetic model. *Proc. Lunar Sci. Conf.* **4**, 481–504.

Warner J.L., Simonds C.H., Phinney W.C., and Gooley R. (1973b) Petrology and genesis of two "igneous" rocks from Apollo 17 (76055 and 77135). *EOS (Trans. Amer. Geophys. Union)* **54**, 620–621.

Warner R.D., Taylor G.J., and Keil K. (1977) Petrology of crystalline matrix breccias from Apollo 17 rake samples. *Proc. Lunar Sci. Conf.* **8**, 1987–2006.

Warren P.H. (1985) The magma ocean concept and lunar evolution. *Ann. Rev. Earth Planet. Sci.* **13**, 201–240.

Warren P.H. (1991) The Moon. *Rev. Geophysics*, Supplement, U.S. National Report IUGG 1987–1990, Amer. Geophys. Union, pp. 282–289.

Warren P.H. and Taylor G.J. (1981) Petrochemical constraints on lateral transport during lunar basin formation. In *Multi-ring Basins*, (P.H. Schultz and R.B. Merrill, editors), *Proc. Lunar Planet. Sci.* **12A**, 149–154.

Warren P.H. and Wasson J.T. (1977) Pristine non-mare rocks and the nature of the lunar crust. *Proc. Lunar Sci. Conf.* **8**, 2215–2235.

Wetherill G.W. (1981) Nature and origin of basin-forming projectiles. In *Multi-ring Basins* (P.H. Schultz and R.B. Merrill, editors), *Proc. Lunar Planet. Sci.* **12A**, 1–18.

Whitaker E.A. (1981) The lunar Procellarum basin. In *Multi-ring Basins* (P.H. Schultz and R.B. Merrill, editors), *Proc. Lunar Planet. Sci.* **12A**, 105–111.

Whitford-Stark J.L. (1981a) Modification of multi-ring basins – the Imbrium model. In *Multi-ring Basins* (P.H. Schultz and R.B. Merrill, editors), *Proc. Lunar Planet. Sci.* **12A**, 113–124.

Whitford-Stark J.L. (1981b) The evolution of the lunar Nectaris multiring basin. *Icarus* **48**, 393–427.

Wichman R.W. and Schultz P.H. (1992) Distribution of lithospheric failure and volcanism in the lunar Crisium basin: Additional signatures of an oblique, multi-ring impact structure. *Lunar Planetary Science* **XXIII**, 1521–1522.

Wilhelms D.E. (1968) Geologic map of the Mare Vaporum quadrangle of the Moon. *U.S. Geol. Survey Map* **I–548**.

Wilhelms D.E. (1970) Summary of lunar stratigraphy-Telescopic observations. *U.S. Geol. Survey Prof. Paper* **599**–F, 47 pp.

Wilhelms D. E. (1972) Reinterpretations of the northern Nectaris basin. In *Apollo 16 Prelim. Sci. Rpt.*, NASA **SP–315**, pp. 29–27 to 29–30.

Wilhelms D.E. (1973) Comparison of martian and lunar multiringed circular basins. *J. Geophys. Res.* **78**, 4084–4095.

Wilhelms D. E. (1976a) Secondary impact craters of lunar basins. *Proc. Lunar Sci. Conf.* **7**, 2883–2901.

Wilhelms D.E. (1976b) Mercurian volcanism questioned. *Icarus* **28**, 551–558.

Wilhelms D.E. (1980a) Stratigraphy of part of the lunar near side. *U.S. Geol. Survey Prof. Paper* **1046**–A, 71 pp.

Wilhelms D.E. (1980b) Geologic map of lunar ringed impact basins. In *Papers Presented to the Conference on Multi-ring Basins: Formation and Evolution*, pp. 115–117. Lunar and Planetary Institute, Houston.

Wilhelms D.E. (1981) Relative ages of lunar basins (II): Serenitatis. In *Reports of Planetary Geology Program-1981*, NASA **TM 84211**, pp. 405–407.

Wilhelms D.E. (1984) Moon. In *The Geology of the Terrestrial Planets* (M.H. Carr, editor), NASA **SP-469**, 106–205.

Wilhelms D.E. (1987) The Geologic history of the Moon. *U.S. Geol. Survey Prof. Paper* **1348**, 302 pp.

Wilhelms D.E. (1990) Geologic mapping. In *Planetary Mapping* (R. Greeley and R.M. Batson, editors), Cambridge University Press, pp. 208–260.

Wilhelms D.E. (1993) *To a Rocky Moon: A Geologist's History of Lunar Exploration*. Univ. of Arizona Press, 477 pp.

Wilhelms D.E. and El-Baz F. (1977) Geologic map of the east side of the Moon. *U.S. Geol. Survey Map* **I–948**.

Wilhelms D.E. and McCauley J.F. (1971) Geologic map of the near side of the Moon. *U.S. Geol. Survey Map* **I–703**.

Wilhelms D.E. and Squyres S.W. (1984) The martian hemispheric dichotomy may be due to a giant impact. *Nature* **309**, 138–140.

Wilhelms D.E., Hodges C.A., and Pike R.J. (1977) Nested crater model of lunar ringed basins. In *Impact and Explosion Cratering* (D.J. Roddy, R.O. Pepin, and R.B. Merrill, editors), pp. 539–562, Pergamon Press, N.Y.

Wilhelms D.E., Howard K.A., and Wilshire, H.G. (1979) Geologic map of the south side of the Moon. *U.S. Geol. Survey Map* **I–1162**.

Wilhelms D.E., Ulrich G.E., Moore H.J., and Hodges C.A. (1980) Emplacement of Apollo 14 and 16 breccias as primary basin ejecta. In *Lunar and Planetary Science* **XI**, pp. 1251–1253. Lunar and Planetary Institute, Houston.

Wilshire H.G. and Jackson E.D. (1972) Petrology and stratigraphy of the Fra Mauro Formation at the Apollo 14 site. *U.S. Geol. Survey Prof. Paper* **785**, 26 pp.

Winzer S.R., Nava D.F., Schuhmann P.J., Lum R.K.L., Schuhmann S., Lindstrom M.M., Lindstrom D.J., and Philpotts J.A. (1977) The Apollo 17 "melt sheet": Chemistry, age and Rb/Sr systematics. *Earth Planet. Sci. Letters* **33**, 389–400.

Wolfe E.W. and Reed V.S. (1976) Geology of the massifs at the Apollo 17 landing site. *J. Res. U.S. Geol. Survey* **4**, 171–180.

Wolfe E.W., Bailey N.G., Lucchitta B.K., Muehlberger W.R., Scott D.H., Sutton R.L., and Wilshire H.G. (1982) The geologic investigation of the Taurus-Littrow Valley: Apollo 17 landing site. *U.S. Geol. Survey Prof. Paper* **1080**, 280 pp.

Wood C.A. (1980) Martian double-ring basins: New observations. *Proc. Lunar Planet. Sci. Conf.* **11**, 2221–2241.

Wood C.A. (1981) Possible terrestrial analogs of Valhalla and other ripple-ring basins. In *Multi-ring Basins* (P.H. Schultz and R.B. Merrill, editors), *Proc. Lunar Planet. Sci.* **12A**, 173–180.

Wood C.A. and Head J.W. (1976) Comparison of impact basins on Mercury, Mars and the Moon. *Proc. Lunar Sci. Conf.* **7**, 3629–3651.

Wood J.A. (1975) The nature and origin of Boulder 1, Station 2, Apollo 17. *The Moon* **14**, 505–517.

Wood J.A., Dickey J.S., Marvin U.B., and Powell B.N. (1970) Lunar anorthosites and a geophysical model of the Moon. *Proc. Apollo 11 Lunar Sci. Conf.*, 965–988.

Zisk S.H., Hodges C.A., Moore H.J., Shorthill R.W., Thompson T.W., Whitaker E.A., and Wilhelms D.E. (1977) The Aristarchus–Harbinger region of the Moon: Surface geology and history from recent remote-sensing observations. *The Moon* **17**, 59–99.

Acknowledgement

Figures 3.11, 3.13, 4.10b and 7.12 are copyright of the American Geophysical Union.

Index

(*Italicized* numbers indicate pages where principal discussion occurs)